CLIMATE REALISM

This book sets forth a new research agenda for climate theory and aesthetics for the age of the Anthropocene. It explores the challenge of representing and conceptualizing climate in the era of climate change.

In the Anthropocene when geologic conditions and processes are primarily shaped by human activity, climate indicates not only atmospheric forces but also the gamut of human activity that shape these forces. It includes the fuels we use, the lifestyles we cultivate, the industrial infrastructures and supply chains we build, and together these point to the possible futures we may encounter. This book demonstrates how every weather event constitutes the climatic forces that are as much social, cultural, and economic as they are environmental, natural, and physical. By foregrounding this fundamental insight, it intervenes in the well-established political and scientific discourses of climate change by identifying and exploring emergent aesthetic practices and the conceptual project of mediating the various forces embedded in climate.

This book is the first to sustain a theoretical and analytical engagement with the category of realism in the context of anthropogenic climate change, to capture climate's capacity to express embedded histories, and to map the formal strategies of representation that have turned climate into cultural content.

Lynn Badia is an Assistant Professor of English at Colorado State University, where she specializes in environmental and energy humanities.

Marija Cetinić is Lecturer in Literary and Cultural Analysis at the University of Amsterdam and a research affiliate at the Amsterdam School for Cultural Analysis (ASCA).

Jeff Diamanti is an Assistant Professor of Environmental Humanities at the University of Amsterdam (Literary and Cultural Analysis & Philosophy).

Routledge Research in the Anthropocene

The *Routledge Research in the Anthropocene Series* offers the first forum for original and innovative research on the epoch and events of the Anthropocene. Titles within the series are empirically and/or theoretically informed and explore a range of dynamic, captivating and highly relevant topics, drawing across the humanities and social sciences in an avowedly interdisciplinary perspective. This series will encourage new theoretical perspectives and highlight ground-breaking interdisciplinary research that reflects the dynamism and vibrancy of current work in this field. The series is aimed at upper-level undergraduates, researchers and research students as well as academics and policy-makers.

Hope and Grief in the Anthropocene
Re-conceptualising Human–Nature Relations
Lesley Head

Releasing the Commons
Edited by Ash Amin and Philip Howell

Climate Change Ethics and the Non-Human World
Edited by Brian G. Henning and Zack Walsh

Involving Anthroponomy in the Anthropocene
On Decoloniality
Jeremy Bendik-Keymer

Resilience in the Anthropocene
Governance and Politics at the End of the World
Edited by David Chandler, Kevin Grove and Stephanie Wakefield

Climate Realism
The Aesthetics of Weather and Atmosphere in the Anthropocene
Edited by Lynn Badia, Marija Cetinić, and Jeff Diamanti

For more information about this series, please visit https://www.routledge.com/Routledge-Research-in-the-Anthropocene/book-series/RRA01

CLIMATE REALISM

The Aesthetics of Weather and Atmosphere in the Anthropocene

Edited by Lynn Badia, Marija Cetinić, and Jeff Diamanti

Routledge
Taylor & Francis Group
LONDON AND NEW YORK

First published 2021
by Routledge
2 Park Square, Milton Park, Abingdon, Oxon OX14 4RN

and by Routledge
52 Vanderbilt Avenue, New York, NY 10017

Routledge is an imprint of the Taylor & Francis Group, an informa business

British Library Cataloguing-in-Publication Data
A catalogue record for this book is available from the British Library

Library of Congress Cataloging-in-Publication Data
A catalog record has been requested for this book

ISBN: 978-1-138-37003-6 (hbk)
ISBN: 978-1-138-37004-3 (pbk)
ISBN: 978-0-429-42828-9 (ebk)

Typeset in Bembo
by Newgen Publishing UK

CONTENTS

FIGURES

CONTRIBUTORS

Philip Aghoghovwia is Senior Lecturer in The Department of English and Cultural Studies at University of the Free State, South Africa. A fellow of African Humanities Program of the American Council of Learned Societies (ACLS), and an NRF Y1-rated scholar, he is completing a manuscript titled "Reading Petroculture in Nigeria's Niger Delta." His most recent essays appeared in *Social Dynamics*, *Safundi,* and in *Fueling Culture: 101 Words for Energy and Environment* (Eds. I Szeman & J. Wenzel. New York: Fordham University Press, 2017).

Lynn Badia is an Assistant Professor of English at Colorado State University, where she specializes in environmental and energy humanities. In 2015 she was a visiting scholar at the University of Cambridge, as part of the Climate Histories Research Group, at the Cambridge Centre for Research in the Arts, Social Sciences, and Humanities. She was a Banting Postdoctoral Fellow (2015–2017) and the Canada Research Chair in Cultural Studies Postdoctoral Fellow (2014–2015) at the University of Alberta, focusing on research in the field of energy humanities. Badia's work is published by interdisciplinary venues, including *American Quarterly*, *Cultural Studies*, *Open Library of the Humanities, and Nineteenth-Century Contexts*, among others. Her book manuscript "Imagining Free Energy: Fantasies, Utopias, and Critiques of America" introduces "free" or unlimited energy as a critical framework for understanding American society since the beginning of the industrial era.

Amanda Boetzkes is Professor of Contemporary Art History and Theory at the University of Guelph. Her research focuses on the aesthetics and ethics of art as these intersect with ecology and visual media of the late twentieth and early twenty-first centuries. She is the author of *Plastic Capitalism: Contemporary Art and the Drive to Waste* (MIT Press, 2019), *The Ethics of Earth Art* (University of Minnesota

Press, 2010), and editor of *Heidegger and the Work of Art History* (Ashgate, 2014). She has published in the journals *South Atlantic Quarterly*; *Afterimage*; *Postmodern Culture*; *e-flux*; and *Antennae: The Journal of Nature and Visual Culture* among others. Recent book chapters appear in *Nervous Systems: Art, Systems and Politics Since the 1960s* (Duke University Press, 2021); *Materialism and the Critique of Energy* (MCM', 2018); *Fueling Culture: 101 Words for Energy and Environment* (Fordham University Press, 2016); *The Edinburgh Companion for Animal Studies* (Edinburgh University Press, 2017); and *Art in the Anthropocene: Encounters Among Politics, Aesthetics, Environments and Epistemologies* (Open Humanities Press, 2015). Her current project, *Ecologicity, Vision and Art for a World to Come* considers modes of visualizing environments with a special focus on the circumpolar North.

Marija Cetinić is on the faculty of the humanities at the University of Amsterdam. Cetinić recently completed a position as assistant professor in the Department of English at York University in Toronto. She held an SSHRC Postdoctoral Fellowship in the Department of English and Film Studies at the University of Alberta. "Signs of Autumn: The Aesthetics of Saturation," her current project, focuses on the concept of saturation and on developing its implications for the relation of contemporary art and aesthetics to political economy. Her essays have appeared in *Mediations*, *Discourse*, and the *European Journal of English Studies*. "House, Library, Field: The Aesthetics of Saturation" appears as a chapter in *Neoliberalism, Value, and Jouissance*, edited by Art, Critique, Theory (Seoul: Booknomad, 2013). Her chapter on "Affect" appears in *A Companion to Critical and Cultural Theory*, edited by Imre Szeman, Sarah Blacker, and Justin Sully (Oxford, UK: John Wiley and Sons, 2017).

Jeff Diamanti is Assistant Professor of Environmental Humanities at the University of Amsterdam. With Imre Szeman, he is the editor of *Energy Cultures: Art and Theory on Oil and Beyond* (West Virginia University Press), and with Amanda Boetzkes, he co-organizes "At the Moraine," an ongoing research project on the political ecology of glacial retreat in Greenland. His first book, *Climate and Capital in the Age of Petroleum: Locating Terminal Landscapes* is forthcoming with Bloomsbury.

Ingrid Diran is an Assistant Professor of English and Environmental Humanities at the University of Wisconsin-Madison.

Anne-Lise François is an associate professor of English and Comparative Literature at UC-Berkeley and the author of *Open Secrets: The Literature of Uncounted Experience* (Stanford University Press, 2008). She is completing a book *Provident Improvisers: Parables of Subsistence in the Time of Enclosures* on the enclosure of seasonal time under industrial capitalism. Essays related to the book have appeared in the collection *Anthropocene Reading* and in the journals *Qui Parle*, *Essays in Romanticism*, *Minnesota Review*, and *Postmodern Culture*.

Barbara Herrnstein Smith is Braxton Craven Professor Emerita of Comparative Literature and English at Duke University. Her publications include *Contingencies of Value: Alternative Perspectives for Critical Theory* (1988), *Belief and Resistance: Dynamics of Contemporary Intellectual Controversy* (1997), *Scandalous Knowledge: Science, Truth and the Human* (2006), and *Natural Reflections: Human Cognition at the Nexus of Science and Religion* (2009). Smith's most recent book is *Practicing Relativism in the Anthropocene: On Science, Belief, and the Humanities* (2018), from which the essay printed here is drawn.

Antoine Traisnel is Assistant Professor of Comparative Literature and English at the University of Michigan. He is the author of Capture: *American Pursuits and the Making of a New Animal Condition* (University of Minnesota, 2020) *Hawthorne: Blasted Allegories* (Aux Forges de Vulcain, 2015), and co-author with Thangam Ravindranathan of *Donner le change: l'impensé animal* (Hermann, 2016). He is also the translator into French of *Sheppard Lee: Written by Himself* (Aux Forges de Vulcain, 2017).

M. Ty is a Fellow at the Institute for Cultural Inquiry, Berlin and an Assistant Professor of English at Clemson University.

Kyle Whyte is Professor and Timnick Chair in the Department of Philosophy and Department of Community Sustainability at Michigan State University. Kyle's research addresses moral and political issues concerning climate policy and Indigenous peoples, the ethics of cooperative relationships between Indigenous peoples and science organizations, and problems of Indigenous justice in public and academic discussions of food sovereignty, environmental justice, and the Anthropocene. He is an enrolled member of the Citizen Potawatomi Nation.

Kathryn Yusoff is Professor of Inhuman Geography at Queen Mary University of London. She recently completed a book, *A Billion Black Anthropocenes or None* (University of Minnesota Press, 2018) that addresses the geology and the raciality of matter. She coedited an SI (with Nigel Clark) on Geosocial Formations in the Anthropocene for *Theory, Culture and Society*. Currently, she is finishing a book, entitled, *Geologic Life: Inhuman Intimacies and the Geophysics of Race* about the politics of inhuman and the white supremacy of matter. The essay is reprinted from *Social Text*.

INTRODUCTION

Lynn Badia, Marija Cetinić and Jeff Diamanti

Realism: an idiosyncratic origin story

In the realm of aesthetics, realism was born the day John Ruskin pulled pathos out from feelings about the weather. The OED's first reference to "realism" in this context is traced to Ruskin's *Modern Painters*, Volume III (1856), as a form of laboring to "[render] the precise detail of the real thing or scene" (Levine 2000, 75). In her review of Ruskin's tome, George Eliot would nominalize this new aesthetic commitment for literature: in her words, "all truth and beauty are to be attained by a humble and faithful study of nature, and not by substituting vague forms, bred by imagination on the mists of feeling, in the place of definite, substantial reality" (Eliot quoted in Levine, 75–76). Reacting against the pathetic fallacy of reading the mood of human history into the character of nature would occasion Ruskin's new term for a representational commitment shared across the disciplines.

Ruskin's realism grew from a set of aesthetic intentions aimed at scientifically observing nature. Caroline Levine argues that Ruskin's realism was not an effort to produce objective mimesis but rather to apprehend nature's *infinite variety*: "infinity is the *sine qua non* of Ruskin's realism" (29). Ruskin insisted that an artist must first resist the mind's rote, even mechanized habits of perception (such as inscribing human emotional life to nonhuman phenomena) in order to see a nature replete in a variety of forms. According to Levine, Ruskin's realism was a radical practice, a labor to perceive the natural world in its ongoing variation of form—aesthetics must be alive to a nature characterized by "changefulness" and its schemes "never to repeat itself." "To form a judgement of the truth of painting," as Ruskin contends in *Modern Painters,* "perhaps the very first thing we should look for ... should be the expression of *infinity* always and everywhere" (Ruskin quoted in Levine, 79).

Is it any wonder, then, that a young Ruskin was captivated by the burgeoning science of meteorology? At the age of 20, Ruskin gave an invited talk at the

Meteorological Society in London, which was published in 1839 as "Remarks on the Present State of Meteorological Science." His opening lines find the infinity of nature to be a common aspect of both knowledge and beauty: "It is indeed a knowledge which must be felt to be, in its very essence, full of the soul of the beautiful … He, whose kingdom is the heaven, can never meet with an uninteresting space,—can never exhaust the phenomena of an hour; he is in a realm of perpetual change,—of eternal motion,—of infinite mystery" (Ruskin [1839] 1903). If Ruskin thrills at this glimpse of what would become a science of dynamic systems, he is equally drawn to the prospect of contributing to the early days of a science that could eventually bring the "autonomous nature" (Merchant 2015) of previous centuries into the purview of precise human perception, measurement, and prediction:

> Times and seasons, and climates, calms and tempests, clouds and winds, whose alternations appear to the inexperienced mind the confused consequences of irregular, indefinite, and accidental causes, arrange themselves before the meteorologist in beautiful succession of undisturbed order, in direct derivation from definite causes; it is for him to trace the path of the tempest round the globe,—to point out the place whence it arose,—to foretell the time of its decline,—to follow the hours around the earth, as she "spins beneath her pyramid of night,"—to feel the pulses of ocean,—to pursue the course of its currents and its changes,—to measure the power, direction, and duration of mysterious and invisible influences, and to assign constant and regular periods to the seed-time and harvest, cold and heat, summer and winter, day and night, which we know shall not cease, till the universe be no more.
>
> *(Ruskin [1839] 1903)*

This is a powerful expression of nineteenth-century scientism as much as it signals the origin of Ruskin's commitment to realism. Still a few decades before the first weather forecasts would instill themselves into the public culture of industrial Britain, Ruskin imagines the microdynamics of globally distributed weather systems of emergence, feedback, and interaction (though he used different terms) as made wholly traceable, decipherable, and predictable through the efficacy of collective scientific observation and measurement. Glimpsing weather as dynamic and complex phenomena—which is to say, evolving phenomena in deep relationality through time—inspired the young Ruskin as much as the possibility of a science able to master it. What Ruskin failed to consider when striking mood from climate, however, were the real material entanglements of planetary forces with human lives, laws, economies, and behaviors. In other words, he never expected to see the uncanny traces of human culture in the theater of planetary weather systems (Farrier 2016)—not the symbolic distortion of reading human emotion onto weather, but dynamically entangled forces, bodies, behaviors, and molecules. Ruskin was able to glimpse the complexity of climate phenomena and make it the founding principle of his aesthetics, but he remained unable to imagine the deep human entanglement with the physical forces of the natural world. So he leapt for something called realism instead of the uncanny.

That same year *Modern Painters* was published, a relatively young Karl Marx—still working on early drafts of what would become *Capital* in the next decade—gave a speech in London, on an occasion marking the dissolution of the Communist League following the failed revolutions of 1848. Marx, ever the enemy of idealism, appears to nevertheless flail into Ruskin's anxiety about the pathetic fallacy, an anxiety about turning nature into a metaphor for the tide of history: "The so-called revolutions of 1848 were but poor incidents [but] beneath the apparently solid surface, they betrayed oceans of liquid matter, only needing expansion to rend into fragments continents of hard rock" (Marx 1969 [1856], 501).

As the edifice of industrial England is figured here as "continents of hard rock," the real force of this tide of history is none other than the "oceans of liquid matter," or the fossil fuels that were already flowing into the rising architectures of industrial modernity. The metaphor is neither mere fancy nor oratory excess—rather, it brings together two sides of the industrial revolution that are figured here as forms of nature and mechanisms of human industry. The force of "oceans of liquid matter" is already prefigured for Marx in the form of "Steam, electricity, and the self-acting mule" (501). Ruskin's and Elliot's calls to study natural forms to service the new aesthetic realism are premised on the renunciation of fancy. In Marx's critical realism, figurative language overflows. But Marx is not studying the same forms as Ruskin and Elliot. The material force of planetarity is prepared for thought by the material force of a startlingly energy-intensive mode of industrial production. The particular features of fossil fuels—the ease with which their energy is stored, transported, and utilized for human projects—enabled the force of the planet to be harnessed to forms of capitalism and colonialism at this particular moment in history, blurring the distinction between natural and historical forms and forces.

The two sides of the planetary metaphor—the natural force of the elements and the mobilized energy of the fossil-fueled industry—burst conjointly into the "atmosphere in which we live," Marx continues, an atmosphere that "weighs upon everyone with a 20,000 lb. force, do you feel it?" (501). From the pathos of weather, we leap now to the historicity of atmosphere. This metaphor imagines history oscillating between oceans of liquid matter, to the factory, and ultimately to the complex dynamics of Ruskin's weather. And if that metaphor stretches the comfort zone of a critical language hostile to idealism, our argument is that figurative mediations of climate will prove as necessary to climate realism today as the scientific facts that mark its reality. The lexicon of natural and harnessed force is ready to hand, here, but it is also already a swerve in the multiple histories that would come to unfold around the realist impulse through the *longue durée* of industrial capitalism and the rapidly changing climate that lurks at the edges of its weather reports, bursting into full-blown anthropocentric climate change once the energy it harnesses for value production returned as the climate of our historical present. The aesthetics of weather, climate, and atmosphere in the Anthropocene—the subtitle of our present volume—is a problem for a realism concerned with the facts of weather, and *Climate Realism* is offered as a reparatory concept that foregrounds the political and ecological contradictions inherent in capital's facility with energy.

Today we know all too well that the fossil fuel industry cannot be represented independently from the political ecology and biophysical realities of climate change, at least not if we are serious about a future disarticulated from the present. And yet, even in our idiosyncratic genealogy of realism—highlighting the historical intimacy of energy, environment, and weather—the aesthetic forms available for mediating the materiality of climate and the historicity that dents it are strained toward figuration. In Amitav Ghosh's landmark critique of literary realism, *The Great Derangement* (2016), it is realism's blush in the face of the weirdness of weather that marks the "great derangement" of contemporary realism: deranged not because it is pathetic in its ecological figurations, but because it avoids them at all costs. Elsewhere we see what Jed Esty calls a veritable boom in realism across media, from reality television to the institutional successes of realism in the literary market—the occasion for which, in Esty's account, is the end of the cold war (323)—yet, realism's return in the post-1989 period for Ghosh only incidentally coincides with the rise of climate science and policy. Ghosh offers a rereading of literary history through the lens of the ecological present, and what he finds "comes as a surprise—a shock, really" that few "of the literary minds of that intensely *engagé* period were alive to the archaic voice whose rumblings, once familiar, had now become inaudible to humanity: that of the earth and its atmosphere" (124). At stake for Ghosh in this surprising rethinking of the intimacy between geo-aesthetics on the one hand and literary history on the other is the strain that comes with attempting to fully represent the volatility of climate change in our present moment using the literary techniques of the realist novel. Departing from those same techniques is just as serious, for Ghosh, since a swerve toward so-called genre fiction is what immediately disqualifies a literary text from the "serious literary fiction" circulated in the *New York Review of Books* and *London Review of Books* (11). Climate change, it would seem, had to soak through minor genres before anyone recognized it as a cultural phenomenon. The resistance to representing climate, in Ghosh's account, is as much an internal problem to literary realism as it is for scientific genres of environmental representation.[1]

Ghosh's book is a critical turning point in the study of the aesthetics of weather, climate, and atmosphere because it historicizes the relationship between the institution of literature and the paradigms developed to explain the physical world across the sciences. Across the different faculties of the university, climate is becoming, among a great many other things, a representational concern, and the simultaneous proliferation of realisms cannot be dismissed as mere coincidence. Speculative, haptic, capitalist, indigenous, petro-magical, perspectival, apocalyptic, and scientific realisms dovetail with object-oriented ontology, new materialism, materialist feminism, and critical realism among others. In our naming of *climate realism*, then, we offer the prospectus that these many turns to the material, to objects, and to entangled and mediated forms of subjectivity can be productively read for and mined as various modes of inhabiting a common context amidst rising ecological awareness—a common condition and effort to grapple with the endless implications of climate change, ecological complexity, and planetary instabilities for thought, politics, ontology, and aesthetics.

The core suggestion of *Climate Realism*, then, is that weird weather today is not weird just because it is unseasonable, but also because it names features of the present that strain the epistemological and historical underpinnings of meteorology, philosophy, realist aesthetics, cultural criticism, and the physical sciences—namely, it erodes traditional distinctions that have stabilized disciplinary work in both the arts and sciences. As Environment Canada's senior climatologist David Phillips suggests, climate change makes weather weird because the climate's changing dynamics more conspicuously domino into and out from insect populations, agriculture, the legacies of colonialism, environmental inequity, financial futures, droughts, and social unrest (CTV, 2018). The realism in our title, then, is not a kind of political realism—a skeptical, realistic, or even resigned realism in the face of climate science projections. Instead, it is our way of noticing that amidst so many other conceptual conjunctures with the term *climate* (scientific, philosophical, speculative, etc.), it might also be one that urgently occasions a bridge between what C.P. Snow (1959) long ago called the "two cultures" of the western university. Climate provides an avenue for thought between the sciences and humanities not because the humanistic disciplines are forced to contend with science, but because the humanities offer a unique view of the intimacy of meaning with the distribution of form. It approaches a realism that must simultaneously admit the fixed coordinates of scientific fact *and* the cultural historicity of those facts.

Realism: the observer, the subject, and the material world

There are, however, a few tensions baked into realism once climate becomes its qualifier. Any climate realism will be more than a little strange because it moves beyond the presuppositions that have shaped its foundation. The realism of climate necessitates moving beyond the God-given right of the self-certain subject doing the thinking, the detailing, and the truth claiming. Traditionally, a realist aesthetic was marked by an accumulation of detail directed at the standpoint of the human observer, so that moving slowly through the weight of the world signaled to a reader that this or that was "real." This focalization of one human observer detailing to another relies on an underlying claim to a type of hermetic objectivity that assumes a clean separation between the observer and the world observed. As Barbara Herrnstein Smith explains in her chapter of this volume,[2] "the objective reality posited in realist and neorealist ontologies is commonly evoked as 'out there,' a realm of Being distinct and apart from the humans who, as it were, just happen to inhabit it" (144). Herrnstein Smith offers, instead, a "dynamic-enactivist" account of cognition and perception, in which the process of living, perception, and world-making are coterminous. It is not only that we shape our environments in the process of living; we, simultaneously, "like other creatures, are continuously shaped—bodily and thus also cognitively—by the features of the niches with which we interact." As

Herrnstein Smith explains, "much of what we experience as external to ourselves and speak of as 'reality' can be understood as the relatively stable, provisionally workable configuration of objects, forces, and processes that we continuously construct through our ongoing, effective-enough interactions with our experiential environments." Our environments/realities "are continuously realized, brought forth and made real for us, through the very processes of our living in them."

If classical realism depended on a human subject to sift through the details, climate realism calls for us to consider that what it means to be a human observer is to already veer toward and with an altered sense of meaning-making, detailing, and also weirding the coherence of the world. Along with Bruno Latour's "nature-culture" (Latour 1993, 7) and Donna Haraway's naturecultures (2003), Anna Tsing (2016) relocates this intermediated expression of climate back into the physical environment of what she terms "third nature" (viii), where the distinction between first nature (or nature as such) and second nature (i.e., socially learned behavior) is dead, scorched, prehistoric: not only are we creatures of carbon, proteins, and water, but PCBs and polymers as well. Our very bodies bear traces of third nature, in our urine samples, fatty tissues, and even our brain structure. Stacey Alaimo's (2010) concept of "trans-corporeality" relocates this intimacy of subject and environment by detailing the porosity of bodies with those that populate the milieu in which they breathe, drink, eat, and absorb each other's toxins (2). We can also look to Elizabeth Povinelli's anthropology of life and nonlife in *Geontologies* and Catherine Malabou's injunction to get psychotrophically addicted to the Anthropocene for recent interventions that help reconceive of both the constitution and capacity of the viewing subject amidst the conditions of a warming world and the stakes of an embodied and immersive analytic for any climate realism.

Third nature, plasticity, subjective entanglement: our "new normals" where the objective observations of Ruskin's realism comingle with the material substrate of Marx's planetary force *in situ*, *in bodies*, *in minds*, and *in history* all at once. While the shifting grounds from which a theory of the ecological subject emerges is not reducible to the pressures of climate change, *Climate Realism* takes these correlated coordinates for an aesthetic more alert to interactive shifts between the bio, cryo, hydro, and geosphere with the earth's atmosphere.

The world changes, our science changes, and so do our cultural and aesthetic modes of detailing that world in the manner of any realism. These emergent frameworks for recognizing the ecological embeddedness of realism's viewing subject also return a multitude of agents to the activity of meaning-making. In other words, the way the Anthropocene calls forth the project of realism requires perspectives that include human, nonhuman, elemental, and even computational semiotes. Take, for instance, experimental filmmaker Ursula Biemann, for whom this new condition occasions an opening onto different ways of understanding the very foundations of media. In her 2015 *Subatlantic*, the freshly thawed microbes frozen before the dawn of the human species reenter the hydrosphere in the North Atlantic and intermix with a new flux of hydrological dynamics. This microbial melt reintegrates with fresh and warming water, CO_2 absorbed from fossil fuel emissions, and the altered chemistry and circulation of ocean currents in a dizzying ecological collapse of past and present time. Biemann sidesteps the melodrama and eschatology of environmental documentary and instead dives into the water and details the strange new condition of this wet ontology, sensed out intimately with video and field recordings that are not representations of reality, but agents within the biosemiotic process. For Biemann, our shifting modes of detailing the world remediate the biosemiotic process as they interacts with it. Charles Sanders Pierce's groundwork for a non-anthropocentric concept of semiotics—in which biology turns into "becoming, learning, and growth" through sign systems (Wheeler 2015, 57)—here plunges into the genetic weirdness of water as the medium of a strange genetic milieu. All living things communicate, remember, and evolve in an ecological milieu, according to Piercian process biology, but this means too that the media devices we use to observe and interact with these sign systems become a part of them. The "realist" understanding, representation, and remediation of these emergent dynamics and beings become operative elements of Anthropocene naturecultures.

Detailing within an ecological frame requires new attunements to distributed and mediated materiality, subjectivity, and perception, as Biemann's work demonstrates. As knowledge about the dynamic interaction of atmospheric processes and human activity has sophisticated, mediating devices and knowledge frameworks have become ever more consequential for representing, modeling, and forecasting the changing climate. As agents in the biosemiotics process and in the cultural mediations of how climate data comes to take on meaning, the technoscientific apparatus is also central to what climate realism takes as the "aesthetics" of weather, climate, and atmosphere. For instance, in Reading, UK the detailing of "nature"—ice core

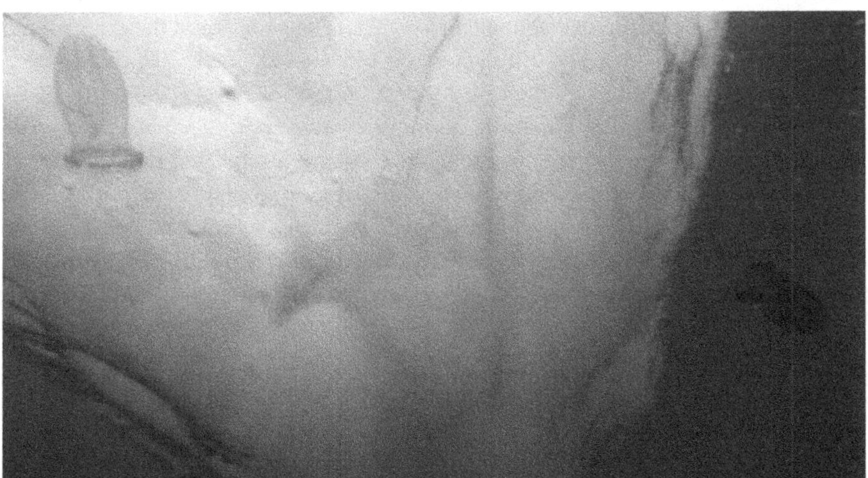

FIGURE I.1 "Ideas rush by like weather events. The art is not to freeze them." © Ursula Biemann, *Subatlantic*, 2015 (quote and image still).

samples, keystone species counts, and bioindicator measurements—seamlessly folds into and out from patent-protected modelling systems created by billion-dollar supercomputers for use by the Royal Netherlands Meteorological Institute to resolve projections regarding the future of Greenland's ice sheet. The media scholar Jennifer Gabrys (2016) calls this "the becoming environmental of computation," where the earth and its atmospheres are integrated cybernetically through sensors, cables, and processors (9). The interactive mediation of globally distributed scientific instruments enlisted to monitor, measure, and mediate the physical world have become indispensable to the clarification of our current and future climatic conditions: indeed, the very concept of global warming is contingent on computational sophistication in the earth sciences. With this in mind, it is obvious that the focalization of environmental detail cannot be reduced to the view of a single type of observer, reader, or temporal horizon that (mythically) characterized Ruskin's era of realism. Both the subject of climate and the temporalities brought into focus by the devices of climate science carry new challenges for the visual, narrative, and discursive representation of climate's full force.

As the flows of information and the crystallization of concepts accrue around the apparatus of climate science, new forms of mediating the strange temporalities, events, and potential futures of a warming world also begin to translate into cultural codes. The reality of climate no longer presents itself as an event rendered for a human view or experience. A hailstorm in August or forest fires in Greenland reference not just the immediacy of the present weather, but also the long history of combustion-based energy, the deep-time process that created fuels uniquely useful for human hands, and a possible future in which the weird weather becomes even more unpredictable and more extreme. In other words, the immediate events of the present have become not only an index to larger dynamic systems but also deep historicity and futurity. Computational monitoring and modeling of earth systems is about both the rendering spatial images of planetary dynamics and the construction of representations of temporal shifts across various metrics like CO_2 concentration, mean temperature, humidity, radiative forcing, and so on.

An important example of how the dissonance between lived experience and the epistemological complexity of climate change data gets mediated into a culturally recognizable form of representation is the time lapse. As a representational scheme for understanding climate change, the time lapse is one attempt to condense and make legible through visualization deep history and futurity. This is why the time lapse has come to matter so centrally in recognizing the multiple scales nominated and threatened by climate change. The aesthetic of acceleration, key to the historical experiences associated with the Great Acceleration, is central, also, for seeing the long view of warming, of atmospheric composition, and of aggregate data that verify the concept of climate in the first place.

The time lapse, with its condensation of data into the details necessary for human interpretation, is also a kind of premise of realism today. When anchored to climate change, the time lapses of history and foreshortened futures bring the technical and elemental, human and molecular, economic and physical into a suggestive relationship—everything from ozone measures, great floods, rates of international tourism, and the world's grand total of McDonald's restaurants are put on a common scale of comparison. Complexity gets rendered into the comparable, and the possible candidates responsible for global warming come into relief. For any climate discourse on the offensive against so many efforts to deny anthropogenic climate change, the data-driven time lapse has been important for representing the pace of environmental change, the entanglement of that change with industrial acceleration, and the inadequacy of solipsistic standpoints committed to what immediately looks and feels true. In environmental documentary, too, from *An Inconvenient Truth* (2006) to *Chasing Ice* (2012) and *Chasing Coral* (2017), the time-lapse image labors to turn climatic expression into a dramatic image of what can otherwise seem too gradual to warrant concern.

Critical efforts to grapple with the strange temporality of global warming, including the slow violence of environmental racism (Nixon 2011), depend on multiple modes of reading. The time-lapse image, while necessarily abstract from the perspective of lived experience, nevertheless helps to foreground the rapid environmental changes that frame the present. The subject to whom climate realism refers as an observer is increasingly placed by ecological dynamics that exceed deictic conventions of address. Time might be said to *lapse* at strange speeds under conditions of industrial acceleration, ecological tipping points, and site-specific legacies of colonialism. As with the time lapse, climate realism should not be read as an attempt to open up to human observation a definitive representational horizon of climate change; instead, it gestures, intermittently and sporadically, in relation to a single point of entry, toward distributed complexity.

To seriously consider the implications of climate realism, then, we must admit that the subject's self-definition is called into question in the act of seeing and

FIGURE I.2 Time-lapse camera, *Chasing Coral* (dir. Jeff Orlowski) 2012.

describing. A second premise of climate realism is that mediated and entangled subjectivities are present to their own redefinition and alternative future meanings while detailing the world. The very concept of climate realism arrives expressively amidst a knotted space of understanding elemental media, the technical apparatus of climate science, and the various interpretants drawn to the scene. So, while photographic, documentary, and literary realism still matter to how climate dents the aesthetic practice of realism, it is in this intermediated space of representations where the arts, humanities, and sciences collaborate on the ongoing challenge to detail climate's history, as well as its present and future truths.

Realism: historicity and futurity

This volume is divided into three sections. Section 1, "The climate of representation," gathers new research on the representational challenge posed by global warming on literary and visual arts. Visual and literary representations of environmental realities today are often characterized by the circulating trope and topography of the damaged landscape, including linear regression graphs in policy and industrial sublimes in landscape photography. The language of risk and remediation soaks the atmosphere of eco-mimetic media, and it codes the timeframe of cause and cure for climate change. Amidst this climate of representation, however, are emergent techniques in the visual and literary arts committed to recomposing the concept, atmosphere, and mood of climate. This section historicizes present representations of climate by reading the long view of its cultural, scientific, and historical development across visual and literary texts, and it aims to distinguish an aesthetics of climate mimesis from climate realism.

Amanda Boetzkes's opening chapter, "Ecological postures for a climate realism," argues that climate realism in conceptual art can be explored to acquire new sensibilities for navigating the conflicting realities of climate, to "ease the spasmic refrain that patterns the battle over the truth about the climate, and resolve it into a re-syntonization of bodies, knowledge, and exchange" (19). Boetzkes considers how art accomplishes this through a negotiation of belief and sensation, of moods and modalities, that work to open up new forms of both perception and embodiment. Boetzkes examines such possibilities in engagements with Rachel Carson's *Silent Spring*, Tony Oursler's video installation *Kepone* (1993), Doug Aitken's *Electric Earth* (1999), and Mary Mattingly's *Wearable Homes* (2004-ongoing).

In his chapter, "Anthropocene arts: Apocalyptic realism and the post-oil imaginary in the Niger Delta," Philip Aghoghovwia convincingly argues that the forward-looking projections of much climate modeling and resilience planning effects an erasure of the *present* apocalyptic realisms: "the apocalyptic future projected as a possible or imminent occurrence is already a quotidian reality in this petroleumscape" (33). In questioning the new emphasis on projections about potential future states in need of present interventions, Aghoghovwia argues that landscapes such as the Niger Delta and Maracaibo in Venezuela "render all scientific projections anachronistic: they are anachronistic and speculative at the same time" (36). As a rejoinder, he calls for texts "to present the reality (depicted in art forms) of climate change and the Anthropocene as a profound actuality of place happening in real time, whose effects are no longer the stuff of projections and possible obviations." In one example, Aghoghovwia reads the experimental video of British-Nigerian artist Zina Saro-Wiwa for a concept of climate rendered through abandoned oil infrastructures in the Niger Delta. In another, Helon Habila's novel *Oil on Water* advances a new materialist poetics drawn from landscapes caught between petroculture and post-oil futures.

Where Boetzkes and Aghoghovwia look for representations of a *situated* climate realism, Anne-Lise François reads the lunar shadows of a warming world for a sense of seasons and the circadian amidst "discrete, determinate, and delicately intercalated units of time" (52). In her literary, historical, and theoretical examples, François finds a counter-poetics to fossil-fueled (and fossil-stored) metrics of "modern clock time," and in turn offers an interdisciplinary mode of reading ecologically for and with elemental rhythms. These three chapters of "The climate of representation" collectively suggest that there is no single disposition of climate realism in the literary and visual arts, but instead offer multiple revisions to the core aesthetic categories and sensibilities that typify realism.

Section 2, "The subject of climate," theorizes modes of subjection and subjectivity unleashed by anthropogenic climate change. Since the turns to the posthuman, material, ecological, and the geological, much theoretical work in the social sciences and humanities has been dedicated to extrapolating a fuller account of our embeddedness with the natural world, or as Dipesh Chakrabarty (2009) has argued, "to bring[ing] together intellectual formations that are somewhat in tension with each other: the planetary and the global; deep and recorded histories; species thinking and critiques of capital" (213). Section 2 offers three accounts by scholars in the fields of literary studies, indigenous philosophy, and human geography. Each chapter considers the ways in which critiques of the subject have prepared us to see climate change as a material condition, and (from the other side) how the new condition of climate smothers the embers of an anthropocentric (and settler colonialist) theory of subjectivity. This section of the book pursues the question: How do we, as Anna Tsing (2015) terms it, practice the "arts of noticing" from within the "interplay of temporal rhythms and scales in the divergent lifeways that gather" (23), especially when we begin to *notice* the multi-species nature of all world-making processes? For Chakrabarty, Tsing, Donna Haraway, Kathryn Yusoff, Bruno Latour, and Timothy Morton, humans are not merely written on by weather, the seasons, and so on; instead, humans share a *geological* subjectivity, as described in Yusoff's chapter here, with nonhuman actors. This section explores the deep consequences of climate change for the human and nonhuman subject, and it makes space for a new theory of the subject(s) of climate.

Kyle Powys Whyte's chapter in this volume argues for an *Indigenous realism* attentive to "relationships between humans and nonhumans, which is a different approach than that of asserting facts and figures or risks of misdistributions of resources" (75). Whyte furthers elaborates on both the historical experiences and philosophical frames that render contemporary climate change into an ongoing process predating the discourse of the Anthropocene. Settler colonialism as such is a protracted and violent *changing* of climate. Ongoing land dispossession and displacement for indigenous communities across North America means that the experience of habitat loss, while phrased as an environmental anxiety in the future tense in the West today, is a painfully old story told by indigenous elders. As Whyte goes on to detail, "Indigenous realism situates today's climate change 'crisis' *systematically* among the different historic and contemporary factors that lead to what becomes labeled 'climate'-related harm and risk" (76). Significantly, however, the historical experience brought into focus by Indigenous realism also helps rephrase the reparative practices occasioned by ecological crisis today. Whyte's stirring conclusion is that "Indigenous realism is about establishing and repairing the relationships of responsibility that can support flourishing human and non-human lives who face constant environmental change" (79).

For Kathryn Yusoff, "The Anthropocene is a name that opens up a speculative dimension to the figurations of planetary thought and material relations. In this speculative space, the very context of matter in which thought exists and takes hold is questioned. The provocation of this new epoch may well be to demand an engagement with geologic realism and a confrontation with the abysmal dimensions of its horizons" (107). In Yusoff's account, the conceptualization of the human as a geological agent named in the Anthropocene requires a fundamental shift in how we conceive of both matter and the materiality of human agency. But just as important to this reconceptualization are the labor and racial histories that mark bodies as more or less proximate to the lithosphere. Hence, for Yusoff, "realism … offers a speculative opportunity in the engagement with geology as a means not just to unearth an anterior posthuman or inhuman position beyond humanism, but as a perspective that could come to terms with both the cosmic potentialities and vicious subjectifications (in terms of race, gender and sexuality) of geologic relations" (102).

As accelerating ecological crises call for interventions, it becomes important to recognize that aesthetic productions of the Anthropocene are often valued in terms of their ability to intervene in the realities most pressing to human survival, adaption, and thriving. As M. Ty argues in her contribution to this volume, modern realisms are subtended by the conceit of the "phantom subject," through which the fantasized absence of the (perceptual, knowing) subject discursively generates the referent of the "real." One reason to name and consolidate a theory of climate realism is to consider how climate aesthetics are haunted by a spectral future in which human well-being and earthy habitability remain possible. "Modern realisms," Ty argues, "formalize and render insuperable the division between experience and knowledge; they generate a vision of the world sundered from the subject's ability to imagine inhabiting it" (94).

The final section of *Climate Realism* collects new work on the philosophical challenge of developing novel realisms adequate to the planetary and historical scales and perspectives invoked by climate change. Human history indexed in the new landscape of climate change turns the very premise of realism into a paradox of beginnings and ends. Does climate realism stage the end of realism as a philosophical and aesthetic desire—since the very relation of subject and object is now verifiably recursive—or the beginning of something new? Analyzing the various traditions of realism that have vied for the truth of representation in the twentieth and twenty-first centuries—such as historical, scientific, speculative, and indigenous realism—the final section of the book asks whether the very basis of realism is canceled in the haze of climate change, or if planetary climate change is the premise of a realism reconciled to climate. Alternatively, does climate change prompt an entire reworking of the questions that traditional realisms have aimed to address?

The futurities under contest across the multiple discourses enlivened and occasioned by climate change all suppose a specific logic of social and ecological reproduction in order to first conceive of that future. In their chapter "The Poetics of Geopower: Climate Change and the Politics of Representation," Ingrid Diran and Antoine Traisnel suggest a new framework for realism:

> we understand realism not as a mimetic replication of a real that exists "out there" and prior to its representation, but rather as a probabilistic regime that defines as real what is predictable and thus, in principle at least, *reproducible*. Realism, in this perspective, conflates representation with reproduction. And what does climate change index if not a crisis in reproduction, when Nature, mythically conceived as that which endlessly reproduces (itself), is found realistically unable to reabsorb human action into its reproductive cycles?
>
> *(123)*

What pressure does this non-reproducibility put on aesthetic modes of representation that seek to see futures never realized and histories able to refine our sense of what is coming?

In the final chapter of this book, "Perplexing Realities: Practicing Relativism in the Anthropocene," Barbara Herrnstein Smith discusses three sites of perplexity relating to climate change. In each of these, she argues, constructivist-pragmatist (often labeled "relativist") accounts of knowledge, science, and belief offer more useful perspectives for conceptualization and practice than classic or current accounts in the tradition of philosophical realism. One perplexity is how to respond to maligned appropriations of important critiques of rationalist-positivist views of scientific knowledge. Bruno Latour's somewhat ambiguous essay, "Why Has Critique Run Out of Steam?" (2004), provides a point of entry. The second perplexity is climate denialism, illuminated by the account of belief systems in Ludwik Fleck's classic *Genesis and Development of a Scientific Fact* (1935) joined with contemporary constructivist-pragmatist understandings of the dynamics of human cognition. The symmetrical methodological principles developed in the sociology of science suggest ways to deal with the third perplexity, which is the existence of multiple, sometimes radically divergent, operative realities. Arlie Russell Hochschild's book *Strangers in Their Own Land* (2016) illustrates the value of such principles. Herrnstein Smith's chapter concludes with an examination of forms of denialism in recent environmentalist writings.

Concluding thoughts: realism in flux

We began by reading a tension wired into the origin story of aesthetic realism in the nineteenth century. There appeared to be two contradictory aims: how, from Ruskin and Elliot's standpoint on the one hand, to study natural forms and processes without relapse into purple prose, and how, from Marx's standpoint on the other, to find the metaphors necessary to phrase the gargantuan force of the industrial mode of production. If the former is a realism measured by descriptive truth, the latter is one instrumentalized toward revolutionary understanding and intervention. The attraction and repulsion between Ruskin's pure facts and Marx's extimate materiality here passes through the sieve of detail and truth and emerges on the other side as a new opposition held in orbit by realism: how to represent the world, and what to do about it. So, if ecomimesis and biosemiotic co-creation are about a realism of the given, what might climate realism have to look like for it to work in the service of an otherwise to the given? The plot thickens, in other words, when the question of how to know climate change is underwritten by another set of questions sensitive to historicity, causality, expression, and *political responsibility*—the pole of realism that swings us back to Marx's historical materialism. The world, after all, did not warm itself.

The birth of aesthetic realism hovered between the industrial revolution and the exacting gaze of meteorology. To what details would Ruskin draw the studious realist's attention, now that the historicity of industry and weather have become inexorably knotted? Now, the observer has to define herself in the process of detailing the world for an uncertain future world she will knowingly shape in the very act of representation. We must simultaneously consider how we define our subjectivity and identity when we designate the material realities that make up our world. We must act with the knowledge that various forms of realist detailing are dented by various types of ethics. All of these inter-relationalities must be consciously held in suspension for the seemingly simple act of detailing.

Before us is dizzying complexity, where historicism has returned but in a most unexpected way—namely, in the very rhythms and flows of nature. Climate, capitalism, and colonialism shift in and out of focus, demanding a deeply difficult and perhaps utterly impossible set of terms around which to articulate a climate realism. Representational strategies, ranging from figures that hold planetarity to a point to sweeping epics that attempt to exhaustively detail the totality of material and cultural entanglements, are weirdly both inadequate and usefully operative. How strange, Bruno Latour (2014) recently commented, that it is human history that appears frozen, and natural history which rages on: foreground and background flipped (13). *Climate Realism* is an effort to wade through this complexity, to potentially generate new but temporary normals, while simultaneously defamiliarizing and demystifying their premises.

Notes

1 For a discussion of Amitav Ghosh in relation to climate realism, see Chapter 8 of this volume by Ingrid Diran and Antoine Traisnel, "The Poetics of Geopower: Climate Change and the Politics of Representation," and Oak Tayler, Jesse. 2018. "The Work of Fiction in an Age of Anthropogenic Climate Change: Review of Amitav Ghosh's 'The Great Derangement,'" *boundary 2,* January 31, 2018. www.boundary2.org/2018/01/jesse-oak-taylor-the-work-of-fiction-in-an-age-of-anthropogenic-climate-change-review-of-amitav-ghoshs-the-great-derangement/

2 See also Barbara Herrnstein Smith, *Scandalous Knowledge: Science, Truth, and the Human* (Duke University Press, 2006).

References

Alaimo, Stacey. 2010. *Bodily Natures: Science, Environment, and the Material Self.* Bloomington, IN: Indiana University Press.

Biemann, Ursula. 2015. "Late Subatlantic. Science Poetry in Times of Global Warming." *L'Internationale Online*, November 15, 2015. www.internationaleonline.org/research/politics_of_life_and_death/45_late_subatlantic_science_poetry_in_times_of_global_warming.

Chakrabarty, Dipesh. 2009. "The Climate of History: Four Theses." *Critical Inquiry* 35, no. 2(Winter): 197–222.

CTV. August 31, 2018. "Take that, Farmer's Almanac: Environment Canada's Dave Phillips predicts milder winter." www.ctvnews.ca/canada/take-that-farmers-almanac-environment-canada-s-dave-phillips-predicts-milder-winter-1.4075532 Accessed Septermber 1, 2018.

Esty, Jed. 2016. "Realism Wars." *Novel: A Forum on Fiction* 49, no. 2: 316–342.

Farrier, David. 2016. "Deep Time's Uncanny Future Is Full of Ghostly Human Traces." *Aeon*, October 31, 2016. https://aeon.co/ideas/deep-time-s-uncanny-future-is-full-of-ghostly-human-traces.

Gabrys, Jennifer. 2016. *Program Earth*. Minneapolis, MN: University of Minnesota Press.

Ghosh, Amitav. 2016. *The Great Derangement: Climate Change and the Unthinkable*. Chicago: University of Chicago Press.

Haraway, Donna J. 2003. *Companion Species Manifesto*. Chicago: University of Chicago Press.

Haraway, Donna J. 2016. *Staying with the Trouble: Making Kin in the Chthulucene*. Durham, NC: Duke University Press.

Herrnstein Smith, Barbara. 2006. *Scandalous Knowledge: Science, Truth, and the Human*. Durham, NC: Duke University Press.

Jameson, Fredric. (1997) 2007. "Reflections in Conclusion." In *Aesthetics and Politics,* edited by Theodor Adorno et al., 196–211. London: Verso.

Latour, Bruno. 1993. *We Have Never Been Modern*. Cambridge, MA: Harvard University Press.

Latour, Bruno. 2014. "Agency at the Time of the Anthropocene." *New Literary History* 45: 1–18.

Levine, Caroline. 2000. "Visual Labor: Ruskin's Radical Realism." *Victorian Literature and Culture* 28, no. 1: 73–86.

Malabou, Catherine. 2017. "The Brain of History, or, the Mentality of the Anthropocene." *The South Atlantic Quarterly* 116, no. 1 (January): 39–53.

Marran, Christine L. 2017. *Ecology Without Culture*. Cambridge, MA: Duke University Press.

Merchant, Carolyn. 2015. *Autonomous Nature: Problems of Prediction and Control from Ancient Times to the Scientific Revolution*. London: Routledge.

Marx, Karl. (1856) 1969. "Speech at Anniversary of the *People's Paper*." *Marx/Engels Selected Works, Volume One*. Moscow: Progress Publishers.

Morton, Timothy. 2013. *Hyperobjects: Philosophy and Ecology after the End of the World*. Minneapolis: University of Minnesota Press.

Morton, Timothy. 2015. *Dark Ecology: For a Logic of Future Coexistence*. New York: Columbia University Press.

Nixon, Rob. 2011. *Slow Violence*. Cambridge, MA: Harvard University Press.

Oak Taylor, Jesse. 2013. "The Novel as Climate Model: Realism and the Greenhouse Effect in *Bleak House*." *Novel: A Forum on Fiction* 46, no. 1: 1–25.

Povinelli, Elizabeth. 2016. *Geontologies: A Requiem to Late Liberalism*. Durham, NC: Duke University Press.

Ruskin, John. (1839) 1903. "Remarks on the Present State of Meteorological Science." In *The Complete Works of John Ruskin*, edited by E. T. Cook and Alexander Wedderburn. New York: Longmans, Green, and Co.

Snow, Charles Percy. 1956. "The Two Cultures." *New Statesman* 6 October 413.

Steffen, Will, Jacques Grinevald, Paul Crutzen, and John McNeill. 2011. "The Anthropocene: Conceptual and Historical Perspectives." *Philosophical Transactions of the Royal Society* 369: 842–867.

Tsing, Anna. 2015. *The Mushroom at the End of the World*. Princeton, NJ: Princeton University Press.

Wheeler, Wendy. 2015. "A Feeling for Life: Biosemiotics, Autopoiesis and the Orders of Discourse." *Anglia* 133, no. 1: 53–68.

SECTION 1

The climate of representation

1

ECOLOGICAL POSTURES FOR A CLIMATE REALISM

Amanda Boetzkes

The ways we imagine and respond to ecological crisis are centrally bound to acts of representation that condition perception. In turn, the honing of vision plays an integral role in shaping the course of ecological knowledge. This aesthetic activity is occurring at a time when scientific knowledge of climate change is hotly contested by corporations, governments, and the general population alike. A constellation of ontological and epistemological demands is putting the "objective knowledge" of the life sciences through its paces. Thus, a heterogeneous climate realism is emerging alongside the ideology of climate change skepticism.

In many respects, the disputes over knowledge-claims about climate change produce a delusional condition. The tension between a popular environmental knowledge and the inhibition of political action produces a cultural spasm, in Félix Guattari's terms: a painful and compulsive mobilization of nervous energies that are both symptomatic of an intensified, excitable discourse and an exploitation of those energies for the preservation of the social body. The question thus becomes: how do we disengage from this refrain of continually "reading signs" of climate change without being discredited as illiterate?

I will argue that contemporary art provides an alternative ground to experience and make claims about the realism of climate change and its impact. I will chart a trajectory that begins with an originary form of ecological denial, the political cover-up, common in the late decades of the twentieth century when governments tested chemicals in depressed cities across North America and then denied the physical effects of their slow violence. This overt denial became an integral facet of the more recent and culturally distributed forms of denial that accompany climate change–related catastrophes. Importantly, I will chart this course through the lens of conceptual artists. Thus, my ambition is not merely to provide an environmental history, but also to show how artists present ecological crisis through alternative sensibilities that attempt to ease the spasmic refrain patterning the battle over the truth about the climate condition, and resolve it into a re-syntonization of bodies, knowledge, and exchange. I consider how art accomplishes this through propositions of moods and modalities that open the possibilities for navigating the new terrain of climate realism. I will examine four *postures*: Rachel Carson's *Silent Spring*; Tony Oursler's video installation, *Kepone* (1993); Doug Aitken's *Electric Earth* (1999); and Mary Mattingly's *Wearable Homes* (2004-ongoing).

Ecological postures for a climate realism

If there is a realism to be thought through the terms of climate, it is not one that is solely defined by scientific data, nor by a hard political line. Rather, it is one that must be considered in connection with *disbelief*. The realism of climate is particular and exemplary; it is a form shaped through the oversights and responsiveness, skepticism, and speculative inquiry that it generates. Climate raises a constellation of ontological and epistemological demands that challenge the ground of objective knowledge of the life sciences while at the same time triggering regressive contestations of the value of knowledge and questioning. Thus, while climate has a shared history with ecology, it cannot be reduced to ecology per se. The drive for a purely instrumental worldview divests the discourse of climate of its hold on critical inquiry of the external world. What a dilemma: one cannot be skeptical without being conservative (i.e., a climate change denier) and one cannot be realistic without being lofty and fantastical, if not paranoid! Climate demands new coordinates of interpretation.

In this chapter, I will link climate realism to acts of representation that express ecological perspective, political scenes, and aesthetic atmosphere. But I will show how these are geared toward a specific climate realism. I do so by running through four postures that I discuss in relation to the ecologist Rachel Carson, installation artist Tony Oursler, video artist Doug Aitkin, and conceptual artist Mary Mattingly. When I speak of postures, I am referring to the etymology of this term, meaning an artificial mental position. These postures enable the transection of a systemic perspective, an aesthetic sensibility, and a critical voice in order to gain purchase of climate from its emergent terms of reality. The four postures are a starting point which give contour to climate realism and its possibilities: (1) a strange stillness, (2) psychogenic spasm, (3) *dancity*, and (4) reflexive carapace. I suggest that it is through the performative gestures of these mental positions toward climate that realism can be thought and figured.

To suggest that climate realism has form and contour is to implicate its presentness and availability to interpretation. But crucially, the interpretation of climate's realism is not of the order of either scientific or political inquiry; it requires an alternative basis for questioning. In his essay, "Various Ways of Questioning About the Thing," Martin Heidegger writes that even to ask the question, "What is a thing?" invokes a modality of thinking and interpreting. He writes,

> The answer to the question "What is a thing?" is … not a proposition but a transformed basic position, or better still and more cautiously, the initial

> transformation of the hitherto existing position toward things, a change of questioning and evaluation, of seeing and deciding; in short, of the being-there (*Da-sein*) in the midst of what is (*inmitten des Seienden*). To determine the changing basic position within the relation to what is, that is the task of an entire historical period.
>
> *(1967, 50)*

Following Heidegger's provocation, I suggest that to probe the question, "What is a climate realism?" is to explore the basic positions that are nested within and emerging from the co-extant discourses of climate and realism as this is being defined in the age of ecology. These basic positions, I suggest, stem from our embeddedness in global ecosystems and the critical reckoning with that embeddedness. That is to say, ecological postures are the transformed basic positions that undertake a questioning, evaluating, perceiving, and acting-toward climate.

Posture 1: a strange stillness (Rachel Carson)

The consciousness of climate stems from ecology insofar as it has developed out of the latter's understanding of its knowledge base as intertwined with systemic crisis. Rachel Carson's seminal 1962 book *Silent Spring* exemplifies this incipient crisis at the origin of ecology. She opens the book with a chapter called "A Fable for Tomorrow," in which she vividly describes an idyllic American town that succumbs to an "invisible curse" which rips through the environment, eliminating life in its wake. Animals, adults, and children fall ill; the vegetation browns and withers; the insects disappear; and there is no bird song. Instead of the plenitude of life, there was "a strange stillness." The spring is silent because a chemical contamination (DDT) has invaded all life right through to its cellular structure, preventing reproduction and system renewal. Carson ends the devastating scene commenting, "No witchcraft, no enemy action had silenced the rebirth of new life in this stricken world. The people had done it themselves" ([1962] 2002, 3).

A perspective of the systemic totality of the town is born of the realization of its contamination. This perspective is produced through its eventful temporal mode: the ecosystem becomes visible at exactly the point that its regenerative organization has been compromised. The value of life is posited belatedly and felt aesthetically as that vitality which has always already slipped through our hands before we even knew what it was. The temporality of belated awareness of the conditions that have overtaken the ecosystem has reached a new scale with the framing of global warming as a climatic event revealing a slowly emerging history. Climate crisis has no focal cause, but rather is the cumulative effect of a slow and insidious progression of industrial history which has already initiated positive feedback loops in global environment patterns. The discourse of climate thus shares with ecology the entangled domain of scientific inquiry, aesthetic sensibility, and political governance. But the appearance of climate *as* crisis means the scattering of the assumed foundations in realism that these discrete disciplines once held. The concept of climate, like ecology, requires a reconsideration of their intimate co-implication with such crises. Moreover, where the logic of crisis as the scene of innovative thinking will strike many as capitalist in nature, we might think of ecology as the domain in which crisis turns on itself to effect a retroactive process of valuation. Retroactive valuation can only take place, however, by acknowledging and enacting its immanence and realism. By the same token, realism can only be confirmed by accepting the retroactive movement of valuation. Carson's posture demonstrates that only by turning an ear to the strange stillness can one become attuned to the sensory wealth of ecology. How, then, does this movement of acknowledgment, attunement, and retroactive valuation happen in relation to climate crisis, in and through its atmosphere of political oversight and aesthetic revealing? And how does it overcome skepticism and other forms of denial?

Posture 2: psychogenic spasm (Tony Oursler)

The ecopsychoanalyst, Joseph Dodds, considers the defense mechanisms against guilt, blame, and responsibility for ecological crisis (2011, 41). He makes an analogy between our responses to climate change and Freud's "borrowed kettle" joke from *Jokes and their Relation to the Unconscious*: a man is told he should replace a pot he has borrowed and returned damaged. He refuses, claiming (1) When I gave it back it was fine, (2) The hole was there when you gave it to me, and (3) I never borrowed it in the first place! In a similar vein, when confronted with overwhelming evidence about climate change and its causes, we can see similar responses at play. First, outright denial: "There's nothing wrong with the climate!" Either it is a conspiracy designed to take away freedom or stop economic growth, or the evidence simply is not conclusive. Second, "There was a hole in the planet when you gave it to me"; the patterns of climate change are natural or it was caused by other people (e.g., India or China)—either way, it is not "my" fault. This is an unconscious displacement of guilt. Finally, "There is nothing we can do about it"; why try? Here, there is an acceptance of the reality, but a depressive paralysis with regards to reparative action.

For Dodds, these behavioral reaction formations are an integral part of more complex positive feedback loops that expedite global warming, such as tropical forest fires, the thaw of tundra permafrost, and accelerated melt of polar ice. The phenomena caused by global warming also intensify the problem because they result in further warming and CO_2 emissions. In other words, defense behaviors are ecological responses in and of themselves, and they aggregate with other phenomena to exacerbate environmental crisis. This interpretation gives a more radical understanding of the unconscious as fully imbricated in the patterns of earthly systems, whether in their balance or their chaotic breakdown. Dodds' theorization of the behavioral defenses against climate crisis rests on the thought that the crisis itself is not subject to a totalizing suppression, but rather is proactively defended against and is therefore incorporated into cultural life through forms of discursive, ideological, and bodily reactivity. Human defense behaviors are supple, and therefore may point directly to an earthly condition, but because of the discrepancy between the lived human world and the scale of objective ecological realities, they are nevertheless exacerbated by the cognitive dissonance produced in trying to reconcile the two. The defense itself may present as a highly ambivalent bodily condition by which an individual incorporates and suppresses the symptoms of environmental crisis.

This oscillation, on the split between acknowledgment and denial of climate crisis, has roots in the history of environmental contamination dating back to the mid-twentieth century, when chemical spills and political cover-ups sprouted up across North America. This time in environmental history is at stake in the early works of American artist Tony Oursler. His video *Kepone* (1993) charts the case of a chemical disaster in the town of Hopewell, Virginia, which was the site of the construction of a factory that produced a deadly pesticide by that name (Figure 1.1). *Kepone* follows from a number of artworks in this time period in which Oursler carried out his concerns with trash and toxic waste as a questioning of the fabric of the mediated image. That is, he grappled with how to imbue the image with a sense of toxicity that would procure the affects of contamination.

The chemical *kepone* (chlordecone) was developed in 1951 by the Allied Chemical Corporation. In 1961, US Food and Drug Administration already knew that it was a toxic compound, and had issued warnings that it needed to be handled

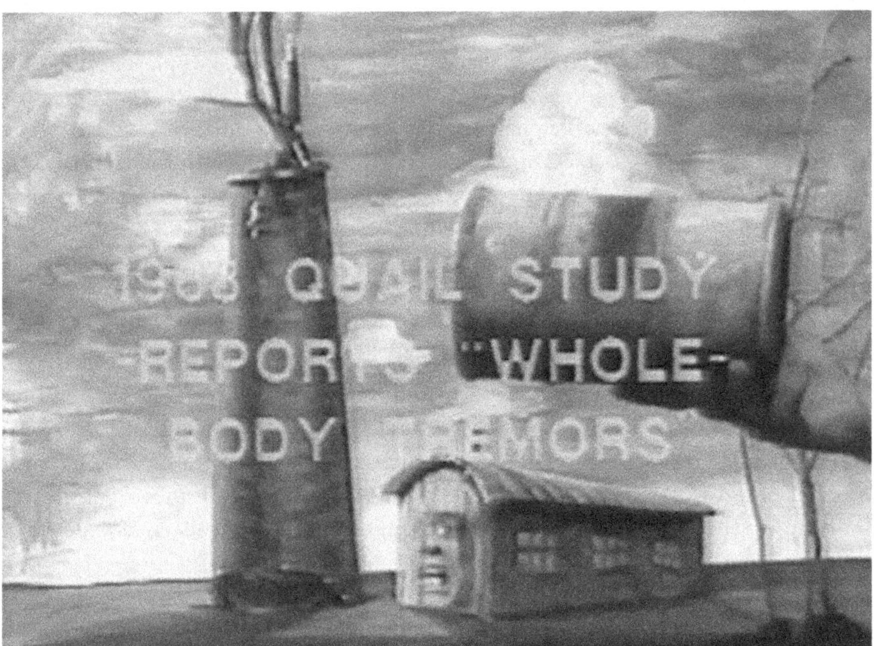

FIGURE 1.1 Tony Oursler, *Kepone*, 1993, video installation. ©Tony Oursler. Courtesy of Lisson Gallery, New York.

with care (Holst and *Encyclopedia Virginia* staff, 2014). However, when the chemical plant was opened in Hopewell in 1966 by a company called Life Science (started by two Allied employees), it seemed there was absolutely no knowledge of kepone's toxicity on the part of management, laborers, or the citizens of Hopewell. Between 1966 and 1975, Life Science dumped its runoff directly into the James River (which was the source of one of the town's main industries—fish and shellfish). Employees suffered the effects of contamination, including brain and liver damage, neurological spasm (uncontrollable shivering and shaking), slurred speech, loss of memory, and erratic eye movement. Employees were never explicitly told that kepone could harm them, so they never wore work gloves and ate on tables covered with kepone dust ("Tragedy in Hopewell"). Only in 1975 after nearly ten years of cover-up were these symptoms investigated, and following that, an inquiry into the effects of river dumping carried out. The factory was shut down in 1975, and a fishing ban was issued for the entire James River, which lasted five years. Allied and Life Science were sued by former workers, residents, and fishermen, for a total of $200 million.

The kepone scandal is the scene of Tony Oursler's representation of the delusional state that arises when a political apparatus negates the physical and affective utterances of bodies that are under the duress of environmental contamination, and creates disbelief by insisting on the priority of employment, industrial progress, and profit. Oursler is known for his video installations that feature "talking heads" projected onto constructions that hinge the internal world of the psyche to the external spatial world. For *Kepone*, Oursler created a textual script that outlined the events of the scandal, created dramatized caricatures of the major players, and then placed these in a primary scene, a painted landscape of the factory, cooling tower, and chemical drum, onto which he projected faces that run through a kind of affectscape of the scandal. The faces speak different perspectives, from the testimony of the factory owners who deny any danger, to those employees who describe their symptoms, those who insist on the importance of employment for their families, and others who describe the symptoms emerging in children. Essentially, Oursler produces a melodrama of the town of Hopewell, with its completely contradictory understandings of the hidden toxin. The faces and musical background are positively campy as they hit the peaks and valleys of both the trauma and the disbelief that the factory poisoned the entire town and its river, not only decimating its own economy *in the name of* economy, but also visualizing the co-extant bodily toxicity and melodrama that afflicts the people. At times, this is humorous like a cult-classic horror film. And yet, that is part of the point; while the work presents itself as a set of internal battles, it is also a forceful representation of a collective experience produced by a documented environmental reality.

Significantly, Oursler visualizes these elements together—an environmental trauma, the physical symptoms of shock, and the affects of skepticism. The disembodied face that speaks to the camera, to itself, to everyone, and no one is a curious blend of self-reflection, bodily utterance, and public address. The worker character, for example, describes the symptoms of neurological distress—uncontrollable shaking—and then being told by the doctor that he needed psychiatric treatment. The hallmark of this political cover-up is the "psychologization" of an inflicted trauma (the claim that the perturbation is imagined and not real). Yet, what compels Oursler is that the victims of kepone are aware of this tactic and yet held in its sway. The cover-up is an integral part of the trauma as the characters become medical subjects of the corporate doctor, political subjects of a town that has been co-opted by a factory, and simply suffering bodies. While knowing something is wrong with them, their distress is unhinged from a known object, and therefore is internalized as an affective climate. Yet the faces (and the eyes in particular) search for an explanatory object—a realism for which to attach a causality that eludes them.

In another installation of that era called *Crypt Craft* (1989), Oursler lays out a general history of chemical catastrophes. He resets the scene of chemical production as a hall of mirrors. In one corner is a dragon's head with a wide-open mouth. From inside the dragon's mouth appears the screened talking head of a man, presumably a victim of chemical testing, who describes his dream of a field of white powder which then turns into a flood. As in much of Oursler's work, the viewer is trained on the movement of the eyes of the subject as they dart from one side to the center in recollection, disclosing the introversion of the person who is talking, and the procedure of scanning one's own thoughts and memories. Set against the psychedelic background, we understand that the account is imagined (dreamed), and yet, given the context of the installation that documents and animates actual cases of chemical poisoning, the dream becomes testimony that is anchored in a broader, nightmarish reality (Figure 1.2). The searching movement of the eyes is directed inside the head, yet the movement is nested within the visual brain, the brain that remembers and fabulates experience of the external world, and which is the site of the realism of ecological poison. Oursler's subjects relive the events of succumbing to an environmental abuse and, through this reliving, give form to Carson's statement "they did it themselves." That is, the reenactment of contamination expresses pain, but that pain is also bound up in the impetus to search the experience further. It is a traumatic repetition that inflicts the painful memory on the subject and the viewer, even as the subject seeks a ground of understanding. The talking heads enact an apparatus in which the subject is systemically contaminated, both externally and through the interpolation of the climate, and then enacts its ecological trauma as a self-contamination in the eternal return of the traumatic scene. Oursler's installation is thus a layout that combines the psyches of the chemists, corporate players, and victims as a positive feedback loop of abuser/abused/self-abuser.

This staging of a collective psyche and its real effects is a climate realism. That is to say, Oursler shows how ecological disasters create an environmental and affective ethos that is shaped and colored by the political, economic, imaginative, and bodily components of a site. While this atmosphere may appear to be "unreal" (imagined, psychedelic, melodramatic, and campy), it is precisely this gathering into ethos that is the terrain from which a realism can be discerned. This is to say that any realism for the planetary condition *must* include the imaginative procedures of the traumatized

FIGURE 1.2 Tony Oursler, *Crypt Craft*, 1989, video installation. ©Tony Oursler. Courtesy of Lisson Gallery, New York.

visual brain: the witness of environmental abuse that is precisely a witness because it has been damaged. These imaginative procedures inform a climate—now more broadly defined to include the human psyche—of the entanglements at stake in such disasters.

Timothy Morton (2013) argues that due to its massive distribution in time and space, climate change exceeds the range of human perception, but that it radiates an appearance. However, Oursler shows how automatic sensory and imaginary processes co-constitute the appearance and perception of such "hyperobjects." In his installations, characters grapple with procedures of representing and reflecting a systemic crisis; they exemplify the form of speculation within the traumatic effects of the crisis. Yet, because this speculation occurs through images, the paradigm of speculation is not exclusively one of scale and perspective but also of interpretation. Oursler lays out the ways such interpretative acts are preempted: the oppressive voices/faces that deny the validity of the complaints, the corporate cover-up, the medical system that explains real symptoms, the overzealous superego that will not permit the individual a thought that it is caught in an apparatus of self-poisoning.

These obstacles are woven into the spasms of the visualizing mind of the victims. Here, I am thinking of Félix Guattari's definition of the spasm: a painful and compulsive mobilization of nervous energies that are both symptomatic of an intensified, excitable discourse and an exploitation of those energies for the preservation of the social body. Spasm is the organism's panic effect as it struggles to interpret the hyperstimulation from the Infosphere (Berardi 2014, 187). We can also think of psychogenic spasm as the juncture between the history of chemical spills and contemporary climate crisis.

Guattari refers to finance capital as a primary example of the Infosphere. But it bears considering that chemistry itself is precisely the development of highly complex information compounds. We might think of the chemical terms by which the "synthetic bonds" that generated the financial crash of 2007 were conceptualized. By the same token, we can consider the informational terms by which petrochemical products such as polymers are rendered ever more complex and dangerous. Where the characters of Oursler's dramas spasm with chemical overload, we may wonder how parallel spasms are taking place in connection with climate conditions, with the understanding that the global atmosphere is subject to aerial dumping from oil production and consumption. Further, the terrains that cross the internal/external worlds of mind and earth can be reimagined precisely as climates. Climate is the coextensive environmental atmosphere and affective ethos that emerges from the systemic interplay between political terrain, scientific development, and cultural navigation that takes us outside of the human frame of reference (or even the frame of reference of life itself). But climate is also environmental atmosphere and affective ethos *as* these occur in bodies, lives, the human psyche, and the planet itself. Climate is the dreamy or delusional state, both internal and external, that makes itself known at the juncture of the spasm. The question becomes, what kind of posture can one take with a brain, a body, and a planet in the throes of a spasm?

Posture 3: dancity (Doug Aitken)

Doug Aitken's video installation *Electric Earth* (1999) advances a posture of coextensive embrace and reactivity to the ecological condition (Figure 1.3). It does so by understanding ecology and the body as intermediated. The video thematizes an earth that is technologically charged in ways that exceed and preclude any restricted understanding of human intentionality. In it, the "electric earth" behaves as an actant that co-opts bodies, objects, and technologies, evoking from them gestures, senses, and atmospheres. The earth is therefore an alien intentionality that makes itself known in its possession and animation of the male protagonist and his surrounding environment.

The video follows a protagonist through an arrangement of liminal zones in a city at dusk. While the setting is urban, it is devoid of people or any distinguishing marker that would identify where it is. Aitken comments that he imagines the protagonist as "the last man on earth," and it is an earth that is between prehistory and posthistory. It is both utopian and dystopian: a wasteland that follows in the wake of human worlds, but which is also anticipatory in that it becomes increasingly animate and thus seems to be waking up, preceding a world to come. The video opens with a scene of a man sitting on his bed watching television. A voice-over narration establishes the protagonist as a climate subject (one that emerges from the climate of the electric earth and who is also subjected to it): "A lot of times I dance so fast that I become what's around me. It's like food for me. I, like, absorb that energy. Absorb the information. It's like I eat it. That's the only now I get." The protagonist gets off his bed and then walks through the otherwise uninhabited city,

FIGURE 1.3 Doug Aitken, *Electric Earth*, video installation with eight channels of video (color, sound), eight projections, four-room architectural environment; installation view at 1999 Venice Biennale. Courtesy of the artist, 303 Gallery, New York; Galerie Eva Presenhuber, Zürich; Victoria Miro Gallery, London; and Regen Projects, Los Angeles. Photo by Doug Aitken Workshop.

absorbing the energies and information of machines that spring to life when he comes into contact with them. He encounters a host of vital technologies that start to radiate and even express themselves: a satellite dish rotates, a streetlight blinks, a security camera searches, a washer in a Laundromat turns, and water pumps at a car wash froth soap. The protagonist is affected by these energies; he absorbs them, one by one, which throws his body into nonsensical gestures. He is pushed and pulled in multiple directions as he incorporates and is incorporated into this assemblage of forces. *Electric Earth* thus carries the viewer through the protagonist's anthropogenesis within an eco-technological climate. Here, the technologies are not instruments or means; they are actants within a global equipment that possesses the protagonist and compels his body with the environment's spatio-temporal rhythms, forces, and ambience.

At the beginning of the video, the protagonist's body works through various spasms precipitated by the forces of the electric earth and eventually finds a rhythm from their impact. His jerky movements resolve into smoother movements and gestures, and the video ends with a scene of him receding down a corridor, swept up in an electric dance. He exists in a protracted state of "dancity" (*dancité*): a bodily output that encompasses the very undecideability between an autonomic reaction and a conscious responsiveness. *Dancity* is Jacques Lacan's term for the capacity of an animal to "pretend" by means of a dance, a lure, or some other choreography (2002, 293). But, as Derrida notes, Lacan's understanding of the animal's dancity keeps it at a pre-symbolic level (2003, 130). Yet, it is an original point of departure into a differentiation between behavioral reaction and reflective response. Lacan imprisons the animal in this distinction, while using it to bolster the structure of the human subject. Derrida puts pressure on this distinction, however, by locating the human subject at this very ambiguity.

My interest here is how the protagonist procures an exuberant dancity or bodily flourishing from within the reactivity of the ecological condition. Like Oursler's subjects, he is taken up with the environment's forces, yet unlocks his body from the corporeal dilemma of the spasm and ushers in a cohabitation with them. Indeed, Aitkin's video concludes with the protagonist's "resyntonization," a calibration of the nervous energies that had bound him up in muscular reactivities. He "ecologizes," to borrow a term from Bruno Latour (2013, 231).

Posture 4: reflexive carapace (Mary Mattingly)

If Doug Aitken stages a resyntonization of bodies in reflexive communication with the assemblage of forces that constitute the earth's ecology, Mary Mattingly provides a bodily carapace that would likewise provide such reflexivity. In her *Wearable Homes* series, Mattingly devises a portable architecture that individuals wear as a superadded layer of the body designed to respond to global climate change (Figure 1.4). The *Wearable Home* encases individual bodies, giving them new perceptual and behavioral capabilities for a volatile planet. In turn, climate is incorporated into the individual movement, habitation, and perspective of the one who wears it.

The *Wearable Homes* project is speculative but takes its terms from the clothing patterns of a cross section of cultural traditions, information technologies, and portable energy systems. Through these components, the architecture of each *Wearable Home* anticipates a subject that is both radically exposed to the climate and who also has the capacity to move through it safely and responsively. Mattingly creates templates derived from Inuit garments, Indian saris, Buddhist robes, Japanese kimonos, and American chains like The Gap, Banana Republic, safari camouflage, and the military in order to construct textiles that protect the body and also provide it with a generalized global form. She relates this generality to an understanding of the clothing's sheltering capacity, insofar as she notes that one wearer would be indistinguishable from the other. However, while this would provide for privacy and anonymity, "the pervasiveness and scrutiny of high-powered networks would still catalog our movements and whereabouts" (Mattingly, n.d.). The home would be outfitted with an information technology mainframe so the wearer could receive and transmit signals, be they via GPS, cell, VA goggles, or the internet. Additionally, the home would be outfitted to inflate in water, with solar panels to provide electricity, warming and cooling fabrics, and batteries that are charged through body motion by power sensor nodes. Each would also have 30 pockets to fit the pills necessary for a month of "mood and health monitoring."

FIGURE 1.4 Mary Mattingly, *New Mobility of Homes, Wearable Homes series*, 2004. © Mary Mattingly. Photo courtesy of the artist.

The *Wearable Home* establishes fictional entanglement of body, climate, and new forms of responsive perception. Its design has a bidirectional trajectory, anticipating the bodily needs of the wearer and the external climate at the intersection of environmental catastrophes, like flooding, the dynamics of power and surveillance, and the culture of communication technologies. With this reflexive second skin, the politics of climate become an implicit sensible dimension of the body's motility, mobility, and expressivity. Mattingly gives this fiction a representation through her photographic series in which she stages the *Wearable Homes* in landscapes that infer the politics of environment that the designs are built for. She photographs lone individuals who stand before landscapes charged with a sense of volatility and impending weather: vast skies with rolling clouds against distant mountain ranges or churning waters and craggy cliffs. The photographs situate the individual specifically in the midst of and *against* climate, standing before it looking outward like the *Rückenfigur* in Caspar David Friedrich's landscapes. The figure mediates a vision of climate and its threats from within the protective and interactive shelter of the *Wearable Home*. The image enacts a perspective of the systemic interplay between body and climate with all the political and technological entanglements of that concept in such a way that climate becomes an integral part of the figure, and in turn the figure is galvanized by the climate.

Conclusion

Rachel Carson's original posture directs attention to a systemic perspective. The strange stillness is the result of a domino effect when an ecosystem is overtaken. But it is also the effect of denying the complex links that bind life systems. The stillness is both the effect and the cause of the problem. Moreover, as Tony Oursler shows, denial produces further distortions under the duress of purely economic logic. Denial can be the symptom of a brain in spasm, a consciousness that is so under an ideological sway that it is unable to cope with the reality of its self-poisoning. It is with this in mind that climate change denial can be rethought through the historical lens of ecology. While there can be no doubt that the disbelief in climate change is delusional, that disbelief is nevertheless the most powerful existing obstacle to perceiving and altering the feedback loops created by climate change. Disbelief is the symptom that has also become (and was always already) the cause of climate toxicity.

An ecological crisis like climate change is enchained to several systems (political, scientific, aesthetic, and planetary) but isolable to none. Locating a singular causality of a system in self-contamination will merely create a spasmodic search loop without object. Yet, the pathways may be eased through the introjection of alternative energies. Such is the realism that the terms of climate suggest—from within the looping of contaminated ecosystems stems an ethos that is colored by the self-reflection of the body and the planet. This ethos demands interpretative gestures to yield its realism, yet it is manifestly present. From the strange stillness, to spasm, to dancity, to the reflexive carapace, the artists discussed in this chapter propose alterations of the coordinates of climate through which its realism must be rethought. For Doug Aitken and Mary Mattingly, the body is resituated in such a way as to propose engagements with the planet that are borne out through technological mediations, dispositions, and forms of speculation pitched toward a responsive climate consciousness. While their postures insist on the entrenchment of the body in the systemic registers that produce climate, they leverage an interplay between the subjective experience and the objective world that moves from spasmodic denial into resolved forms of embodiment. These postures are thus the starting point to accept the task of rethinking climate's realism.

References

Berardi, Franco "Bifo". 2014. "Chaosmic Spasm and the Educational Chaoide." In *Deleuze and Guattari, Politics and Education: For a People Yet to Come*, edited by Matthew Carlinand Jason Wallin, 181–194. London: Bloomsbury.

Carson, Rachel. (1962) 2002. *Silent Spring*. New York: First Mariner Books.

Derrida, Jacques. 2003. "And say the animal responded?" *Zoontologies: The Question of the Animal* edited by Cary Wolfe, 121–146. Minneapolis: University of Minnesota Press.

Dodds, Joseph. 2011. *Psychoanalysis and Ecology at the Edge of Chaos: Complexity Theory, Deleuze/Guattari and Psychoanalysis for a Climate in Crisis.* London: Routledge.

Guattari, Félix. 1995. *Chaosmosis: An Ethico-Aesthetic Paradigm*. Bloomington: Indiana University Press.

Heidegger, Martin. 1967. "Various Ways of Questioning About the Thing." In *What is a Thing?*, 1–54. Chicago: Henry Regnery.

Holst, A.M. and *Encyclopedia Virginia* staff. 2014. "Kepone (Chlordecone)." In *Encyclopedia Virginia*. Last modified May 16, 2014. Accessed June 13, 2017. www.encyclopediavirginia.org/Kepone.

Lacan, Jacques. 2002. "The Subversion of the Subject and the Dialectic of Desire in the Freudian Unconscious." In *Écrits*, translated by Bruce Fink. New York: W.W. Norton & Company Inc.

Latour, Bruno. 2013. *An Inquiry into Modes of Existence: An Anthropology of the Moderns*, translated by Catherine Porter. Cambridge, MA: Harvard University Press.

Mary Mattingly. n.d. "Wearable Homes." Accessed June 19, 2017. www.marymattingly.com/html/MATTINGLYWearableHomes.html.

Morton, Timothy. 2013. *Hyperobjects: Philosophy and Ecology After the End of the World*. Minneapolis: University of Minnesota Press.

"Tragedy in Hopewell". 1976. *Time*, February 2, 1976.

2

ANTHROPOCENE ARTS

Apocalyptic realism and the post-oil imaginary in the Niger Delta

Philip Aghoghovwia

Introduction

Prevailing iterations of the Anthropocene tend to elide the specificities of site-based, local events of climate change, presenting human-induced changes of the earth's climate in abstract terms by means of scientific modeling and speculative projections. This chapter analyzes recent Niger Delta post-oil envisioning that enables us to approach the Anthropocene as a profound actuality of place. This landscape evidences devastations of apocalyptic proportions that render all scientific projections anachronistic; the apocalyptic future projected as a possible or imminent occurrence is already a quotidian reality in this petroleumscape. It is these textures of the real and the concrete that constitute what I call apocalyptic realism in the selected texts: the work of Zina Saro-Wiwa and Helon Habila's novel *Oil on Water*, which I analyze later. Thinking of oil energy sites in relation to the Anthropocene involves thinking the environments where humanity's voracious overreach has been most telling and visible. The artists and writers I investigate here evince how the earth, nature, and all possibilities of life crumble under the sign of oil, bearing witness to the series of events that mark the culmination of the world's end. Yet at the same time, as I intend to argue, thinking new regimes of energy requires a deliberate attempt to begin from the rubble of the incumbent regime of oil and to envision a world beyond oil, reading apocalyptic realism as an organizing trope by which anthropogenic climate change is apprehended.

The term "Anthropocene" is credited to the Nobel-winning chemist Paul Crutzen and former biologist Eugene Stoermer (2000) in a scientific newsletter, published at the turn of the twenty-first century. According to Crutzen and Stoermer (2000), Anthropocene tentatively names the current epoch of the earth's ecology in a time when an otherwise infinitesimal human history, within the broader natural history of the earth, has become an agential character, a "dramatis personae," and scripter of the "geostory" of our blue planet (Altvater 2016). They enumerate with astounding precision human activities on the planet in the last two centuries, beginning, for instance, with James Watts's invention of the steam engine in 1784. These lead them to consider the impact of humankind on the earth's ecology as constituting "a significant geological, morphological force" proportionate to an event of nature that brought about other geological epochs or time scales, such as the Holocene, which has lasted for 11,000 years (Crutzen and Stoermer 2000, 17; Zalasiewicz et al. 2010); hence the new term "Anthropocene," denoting the anthropogenic character of the earth's changes.

Since the publication of that short essay, there have been intense debates regarding the various implications of this coinage (see, inter alia, Crutzen 2002; Steffen 2010; Chakrabarty 2013; Crist 2016), one being over the date that best marks the onset of the Anthropocene (Crutzen & Steffen 2003; Altvater 2016; Chakrabarty 2017). Since 2017 the World Geological Society has agreed on the date of 1945, set from around the end of World War II, because it marked the beginning of the Great Acceleration, the explosion of the atom bombs, and the intensification of nuclear activities by the competing superpowers of the twentieth century (Steffen 2013; Morton 2013; Chomsky 2018). Still, no concrete agreement has been reached on the extant implications of the term; no one has yet had the final word. In short, the term Anthropocene has raised more questions than answers, and some critics have begun to call the sobriquet into question.

Eileen Crist (2016), for example, questions the "narcissistic overtones" that the "naming of an epoch after ourselves" carries (24), noting that such a gesture is self-reflexively "anthropocentric" and mutually coextensive with the "worldview that generated the Anthropocene … in the first place" (14). Other thinkers would even like to see a framework that situates the anthropogenic effects of the earth's ecology within a critique of capitalism. Jason Moore's edited collection, with its pithily suggestive title *Capitalocene* (2016), is an important intervention in that regard. Similarly, Naomi Klein, in her bestselling book *This Changes Everything* (2015), suggests the climate crisis is attendant on the prevailing political economy of capitalism and therefore a human problem that needs an equally human solution, but that "we have not done the things that are necessary to lower carbon emissions because those things fundamentally conflict with deregulated capitalism" (18). Klein surmises that the climate crisis is "a battle between capitalism and the planet" (22). Thus, any genuine attempt to address the problem would have to confront head-on the prevailing political economy of capitalism. Dipesh Chakrabarty (2017) presents a somewhat different view, suggesting that a critique of capital in and of itself is "insufficient to account" for the processes of the earth's changes, as these require "geological and paleoclimatological knowledge" in addition (41). He notes that what specialists need to reflect on, however, is the "uneven responsibility" it places on the generic term "man"—the *Anthropos* being attributed with a geological force, a fact, he notes, that needs to be addressed within a framework of justice (42).

To wit, the interminable wrangling over the Anthropocene concept is attendant on its unevenness (Chakrabarty 2017; Nixon 2017; Klein 2014, 1–28) and its universalizing tendency. This universalizing discourse of the Anthropocene ramifies in at least two ways. First is the entity *humanity*—or *Anthropos*—around which (or more precisely, around whom) the Anthropocene discourse is organized. It is a term much despised, noted Chakrabarty, because of its unresolved historical baggage and "ideological trappings" (42). Take for instance the central thesis of the Anthropocene, which suggests that human activity is fundamentally altering and hastening the earth's climate and weather patterns. The implication of this notion, despite the various qualifications proffered by scholars cited above, is that all of humanity is to blame for the voracious exploitation of the earth's resources. Yet, it is a crime, I would agree, committed by a few industrialized countries, the burden of whose consequences is being distributed and democratized amongst all nations, including the poor in the Global South. The science and politics of global warming identify *collective humanity*—as the term *Anthropocene* suggests (Crutzen & Stoermer 2000, 17; Crutzen 2002; Steffen 2010)—as a geological force driving the earth's ecology and rewriting its natural history. This otherwise convincing discourse presents humanity as if, historically, the label has always meant a unified constituency of human beings; as if humanity was not only until recently bifurcated between what Jean-Paul Sartre describes as "men" and "natives" (1961, 7). The first were the colonizers and empire-builders who bestrode the earth and exploited its resources, including the "natives" who were attributed with an "otherness" that turned them into equally exploitable resources, just like the earth itself. The other humans were, at best, spectators to the drama of the "men's" voracious onslaught on the earth—that is, when they were not part of the earth's exploited resources themselves. Indeed, as Rob Nixon notes, "homo sapiens may constitute a singular actor [in the Anthropocene] … but it is not a unitary one" (2017, 44). What the science and politics of global warming have not sufficiently accounted for is that the "humans" it refers to are just a few concentrated in one small, disproportionately rich, part of the world. These humans, who consist of less than twenty percent of the world's population or "one-fifth of humanity" (Chakrabarty 2013; Chomsky 2018) at any given moment in history, are citizens of the industrialized nations whose rapid development was driven by the easily accessible and abundant supply of carbon-based energy. The citizens of these industrialized nations, as Timothy Mitchell (2009, 400) notes, have developed unsustainable lifestyles and profligate cultures, consuming 80 percent of the earth's resources, and in the process putting "carbon that was previously stored underground" in the atmosphere, "where it is causing increases in global temperatures" and putting great strains on the earth's capacity to recalibrate itself.

The second ramification of the prevailing iterations of the Anthropocene is its tendency to elide the specificities of site-based, local events of the earth's changes. This is the focus of my own reflection in this chapter. The changes in the earth's climate are presented as characterized by unusual, volatile weather patterns, often abstracted and charted by means of scientific modeling,[1] the devastating effects palpable and also imminent, dire consequences which could still be staved off and possibly "reversed to the Holocene-like stage" if certain steps are taken (Steffen 2013).

In this chapter, I argue that there are examples of actual events that do not fit easily within this logic, precisely because there are landscapes laid waste by human activities such as we see in places of oil extraction like Baku in Azerbaijan, Maracaibo in Venezuela, or, more concerning, in Nigeria's Niger Delta.

The sites of oil extraction—specifically the defunct ones in these images—present startling evidence of the wide divide between humanity's march toward progress and material abundance and the landscapes of abandonment that follow in the wake of humanity's onward trajectory. These landscapes present devastations of apocalyptic proportions that render all scientific projections anachronistic: they are anachronistic and speculative at the same time, as a consequence of the particular mode of rereading history and of anticipating a future toward which the historicity of the present leads us. The apocalyptic future projected as a possible or imminent occurrence is already a quotidian reality in regions that seldom feature in the global climate discourse. Their devastation is no longer the stuff of any possible obviations; the apocalypse is here to stay. Voracious extraction has produced dire conditions that are irreversible. Witness for instance the polluted rivers, farmlands laid waste, and several villages across the marshlands and creeks of the Niger Delta abandoned because they can no longer support any form of life. This transformed landscape of extraction has forced local inhabitants to reconfigure the political economy that governs their existence, precisely because the old system is no longer consistent with what is possible in the face of the extraordinary ecological condition that is the Anthropocene.

FIGURE 2.1 Still frame from Basil Gelpke and Ray McCormack, dir. 2006. *A Crude Awakening: The Oil Crash*. Lava Productions AG. DVD, 83 minutes.

FIGURE 2.2 Still frame from Basil Gelpke and Ray McCormack, dir. 2006. *A Crude Awakening: The Oil Crash*. Lava Productions AG. DVD, 83 minutes.

FIGURE 2.3 Image of a Niger Delta community in the wake of oil extraction and the local insurgency that constitutes itself as resistance group. (AFP, *the Economist*).

Thinking of oil energy sites in relation to the Anthropocene involves thinking the environments where humanity's voracious overreach has been most telling and visible. These everyday environments of fossil fuel extraction offer conditions so dismal that they force upon one a sense of an apocalypse, compelling the individual to speak in the idiom of catastrophes and disasters, bearing witness, in eschatological terms, to the possible endings and exterminations, and to reach for horizons that would bring about new regimes of ontology. What new ways of conceiving culture, or, in fact, ontology, are representations of a post-oil environment offering in the light of the Anthropocene? What kind of historicity does this quotidian environment possess (or offer) in the context of the Anthropocene? It is these textures of the real and the concrete that constitute what I call apocalyptic realism in the selected texts that I analyze here.

I arrange these texts into two genres—short performance art captured in images and the written novel form—in which attempts to move away from oil and its troubled world of environmental devastation, while being (re)configured and scripted, also inflect the Anthropocene in all its geopolitical and quotidian scales. For instance, in Zina's art, the performances of masquerades captured in photographs are done at the sites, as well as on top, of a decommissioned oil wellhead (Z. Saro-Wiwa 2015). This is a powerful transgression of an otherwise restricted space, symbolizing the end of the oil regime and a new beginning in which the previous space of extraction is reconstituted into a place of quotidian existence, not only of the human but also of the extra- and nonhuman, and of testing new boundaries of ontology. In Helon Habila's novel *Oil on Water* (2010), the analogical connotation of the oil's end takes a more spectacular and grisly form. The text foregrounds three types of place: landscapes laid waste by oil extraction and its aftermath of abandonment; hastily evacuated villages "as if a deadly epidemic had swept through" them (4); and sites of contaminated waterbodies and decomposing animals in the wake of a strafing of fishing communities by state military forces and local militants, at war over control of the oil. All these index the Anthropocene in Nigeria's Niger Delta as an imprint of oil. The novel tropes on the catastrophic and the eschatological, both of the local environment and the human population that live in proximity to the extraction sites. Yet, what emerges out of this intimacy of violence is the unprecedented resilience of a religious and fishing community on Irikefe Island. Determined to outlive the upheavals around it, the community survives its ruins of oil wars, is rebuilt and repeopled by equally resilient individuals, such as Boma, Gloria, and the Head Priest Naman, who carry the visible scars of the oil violence.

In short, the texts conceive new ways of being, ones that are *beyond* the ken of modernity that the oil ontology inscribes in the Niger Delta. I use the term *beyond* precisely because the new world as imagined in these works emerges out of the aftermath of the one that attends oil extraction; the texts do not eschew the usual ontologies of the real: environmental catastrophes, violence, and other deprivations that mark the oil encounter in the Niger Delta. Rather, in foregrounding these negations, their harsh realities amplify and climax in an apocalyptic cadence. Nevertheless, out of this apocalyptic topos emerges a turning point in which an alternative world of oil's banishment comes into being. The writers and artists I here investigate suggest that oil extraction has wrought destructions of apocalyptic proportions on the environment; thus, their work is an envisioning of a new way of regenerating their small piece of Planet Earth in a manner that banishes oil from this reenvisioned world. Yet, in doing so they begin by first accounting for the extant ways in which the world under the current regime of oil energy has come to a conclusive end. The collective vision evinces how the earth, nature, and all possibilities of life crumble under the sign of oil, bearing witness to the series of events that mark the culmination of the world's end. Yet at the same time, as I argue, thinking new regimes of energy requires a deliberate attempt to begin from the rubble of the incumbent regime of oil and to envision a world beyond oil, reading apocalyptic realism as an organizing trope by which anthropogenic climate change is apprehended.

Playful arts of the everyday: using cultural Idiosyncrasy to transgress the global

The first set of representational forms that I want to look at is the work of British–Nigerian artist Zina Saro-Wiwa. Zina is a video artist, photographer, and filmmaker. She is the daughter of the slain environmental activist Ken Saro-Wiwa. Her work features video installations, photographs, and sound installations that stage performances related to everyday living, cultures, customs and traditions, and the environments of the Niger Delta, specifically the Ogoni. On her website, it states that her installations "use folklore, masquerade traditions, religious practices, food and Nigerian popular aesthetics to test art's capacity to transform and to envision new concepts of environment and environmentalism" (Z. Saro-Wiwa 2015). At center stage in these eclectic performances is the defunct globalized infrastructure of oil extraction symbolized by the oil wellhead, which, in Nigeria's oil industry, is ironically (and cynically) called the "Christmas Tree"—meaning an object of beauty that bears gifts and elicits joy in all who behold. The trajectory of oil extraction in this place has left in its trail anything but gifts, and elicited no joy of any sort in the communities that live in proximity to the oil installations, as witnessed, for instance, by the Ogoni affair, a subject that has attracted much scholarly attention.[2] Elsewhere I myself have written about the profound paradox inherent in the bioregion of the Niger Delta, of which Ogoniland is part (Aghoghovwia 2017b). What this landscape evinces is the very *speech act* of paradox: a site with prodigious oil reserves and overwhelming social and ecological deprivations, a situation that at once elicits a rich body of environmental writing and fashions a culture of militancy whose politics of resistance exacerbates the issues it purports to address (241). One figure who embodies this incongruous relation of oil extraction, nonviolent resistance, and violence is Zina's famous father, Ken Saro-Wiwa. In his lifetime, he succeeded in challenging the state's authority over oil revenue and environmental despoliation through non-violent strategies of cultural protests and imaginative writing with remarkable effects. However, he met with a tragic death at the hands of the government in the cause of his non-violent methods.[3]

Perhaps categorizing the paradox of the "Christmas Tree" as a *metaphor* of lived reality does not do complete justice to the workings of paradox in this oilscape. Paradox bears out in insidious and tangible ways in the Niger Delta.

The Christmas tree wellhead in Zina's art is presented, on the one hand, as a monument to human hubris, consigned to the past, never to be fetishized again. On the other hand, it is also a heuristic device for thinking the emergence of a new

FIGURE 2.4 Still from Zina Saro-Wiwa, dir., 2015. *Karikpo Pipeline,* 5-channel digital video, 27 minutes.

ontology in which oil extraction exhausts its purchase, discontinued and consigned to history, and in which the locality of its extraction is reclaimed and reconstituted into a place of being. The site of the wellhead, which used to be a territorialized space of enclosure, protected and enclaved, becomes instead a playground for the local community: a place where they can lose themselves in a new sense of being, a sense of owning oneself and of reinventing ways of being local and of reestablishing relationship with one's culture. In fact, the site of the decommissioned wellhead becomes one for experimenting with a new lease of being, of a nonhuman-centered sociality, of being able to commune with the environment, in a world where deities, spirits, animals, humans, and other cosmic beings can constellate without the intrusion of the disruptive modernity that oil extraction perpetuated in that locality.

Dipesh Chakrabarty has noted that "postcolonial studies at this historical juncture must have to confront two imperatives of our time: globalization and global warming" (2012, 1). Thus, Zina is de-globalizing the landscape of oil extraction in Ogoniland. Her art is playful but it is consciously organized to rehabilitate and reinhabit the landscape, which, hitherto, had been given to oil extraction. The local landscape, once demarcated for the exclusive purpose of oil extraction, is being repopulated with the presence of mythical creatures, deities, ancestral masquerades, and the human, all constellating around an object from which they were previously completely territorialized. In his widely cited essay, "At the Edge of the World: Boundaries, Territoriality, and Sovereignty in Africa," Achille Mbembe provides a fascinating analysis of how territorial formations in Africa domesticate globalization into what he calls, drawing on Heidegger and Fernand Braudel, "world time" (2000, 260). Mbembe discusses this world time as the peculiar configuration of space and time in the operations of globalization in Africa, organized to control spaces and resources, and, in the process, "proceeds in the material deconstruction of existing territorial framework," excising "conventional boundaries" and simultaneously creating "spaces of enclosure intended to limit the mobility" of local populations adjudged "superfluous" (284). It is this notion of local superfluity that Zina's art is challenging by engaging the territoriality of oil extraction in a playful act of occupation, bringing their excess of being into the very stage where the Anthropocene is most visible: the globalized infrastructure of oil in the Niger Delta.

Zina renders the symbol of that globalized infrastructure, the oil wellhead, provincial by suffusing it with local characters and turning its surrounding plain into a playground for these beings. I read this as a powerful mode of transgressing the global and of envisioning a way out of what Tim Morton in a different context calls "object-oriented ontology" (2013, 1), which I claim to be a consequence of oil energy and the fossil-fueled modernity it brings into being, out of which there appears to be no alternative. Crucially, I hasten to add that my interest in object-oriented ontology (also known as OOO) is not necessarily in line with Morton's own, Kantian, reflection, precisely because of the speculative realism that constitutes his philosophical analysis, interested in the intrinsic albeit transcendental traits of objects that are neither material nor locally apprehended (1). This is not very useful for my own reflection here. On the contrary, OOO extends speculative realism into the material dimensions of existence. Precisely because objects, beyond their ken of transcendence and realist speculations, do limn in excess of their so-called non-materiality and, therefore, become thoroughly material, or become apprehended as such in instances where intimately apprehended. OOO can in this context be thought of as allegorical of the object of oil, its ontology in the Niger Delta; the particular mode of its extraction is undertaken, as I have argued elsewhere, within a form of globalism that "substitutes local subjectivity for pluralistic non-being" (Aghoghovwia 2017a, 35). Thus, my interest is in a work of art that engages a mode of being that orients itself, materially, toward (or more precisely, against) a particular object signified in various forms, in this case the wellhead. Zina's is an attempt to reconfigure new possibilities of life that decenters the one built around fossil fuel, unconditioned by instrumentality and thoroughly idiosyncratic as is often the case with new beginnings.

In her attempt to envision a new beginning beyond the world of fossil-fueled modernity, Zina has artfully domesticated a space previously internationalized, reconfiguring the landscape as a space ontologically charged with the quotidian, a place where one could play and be at ease. Notice also the mask worn by the performer. The mask is a representation of a local deity of the Ogoni people, possibly no longer in fashion, or at least not since after those dark days of their struggles with the Nigerian state. Zina is using cultural artifacts, such as masquerades, to challenge the form of globalism by which the international infrastructure of oil extraction operates. Indeed, the image of the masquerade as depicted in Zina's art is not new. The masquerade is both a material image of culture as well as metaphorical in nature. Zina's father Ken Saro-Wiwa, in his time, had also employed the image of the masquerade, to great effect, in his creative work. In his polemic *A Month and A Day* (1995), he describes the comprador petty bourgeoisie as "black colonialists who wear the mask of Nigerianism," who in truth are "masquerades leashed to a rope held by an unseen hand, and steadied by the oil of the Ogoni and other peoples in the Niger Delta" (K. Saro-Wiwa 1995, 126; also quoted in Caminero-Santangelo 2014). So, perhaps reclaiming and reappropriating the image of the masquerade into an agent for good is a useful way of signaling the end of the oil regime and of generating a productive future out of that terrible history.

In Zina's art, such cultural artifacts are rehabilitated and restored to, once again, feature at the center of cultural life, but this time in ways more positive and generative. What we find, therefore, is a simultaneous constellation of the past, the present, and the future. Her art constitutes a drive towards a future imagined in a post-extraction society—also a reinscription of the past that precedes oil extraction. What the artist appears to suggest is that any attempt to think a way out of the current regime of oil extraction and dependence must not be divorced from the place of knowing, the place of the quotidian, of culture where one is grounded both constitutionally and historically. Therefore, her art is a performance that comes with a caveat, which is that an attempt to reach for the new must not yield completely to this new—the abstract or the ostensibly scientific—but must bring into the fray something valuable from the past as a means of reconstituting the collective future.

Assembling human and nonhuman characters around the decommissioned oil wellhead is indexical of an awakening to a new regime of vision; a vision alert to the limitations of the human who, in his/her sterile politics, often pretends to rational and scientific solutions to global warming where none is afoot. This is why Zina tropes on the otherwise nonsensical and the playful as a necessary starting point from which to imagine new horizons of possibility. It is also an attempt to recuperate and reconfigure this particular space as a place of everyday, commonplace existence for all beings rather than a fetishized space of commodity extraction.

Of nondescript lives and landscapes: blood, scar, and resilience in Habila's *Oil on Water*

Helon Habila's novel is also relatively uninterested in both the grand debates of global warming and of capitalism's myriad defects, including its atrocious impacts on the environment. Rather, *Oil on Water* articulates in cinematographic form the ruins of local environments and peoples whose existence unfolds in proximity to oil extraction. Reading scenes in the novel is grisly; it is an excursion into a world in the aftermath of an apocalypse wrought by oil wars in the locality, ones of sheer brigandage and reckless rivalry:

> But the soldiers came early the next morning. First they came in a boat, and there were only five of them. They were on routine patrol; they hadn't

> known the militants were here, and they ran into ambush—it was a massacre. They were all killed, instantly. The militants had machine guns and grenades. It was awful. But the soldiers must have called for back-up because this morning the helicopter came and started shooting at everything beneath it, indiscriminately. People running and jumping into the water. It was awful. Awful. The water turned red. Blood, it was blood. But in the confusion the rebels slipped away and left the villagers to face the soldiers.
>
> *(Habila 2010, 154)*

It is a tale of the militarized sociality of oil extraction in the Niger Delta, a sociality played out in minute details of local quotidian life, their everyday misery featuring as the very "speech act" of those abstractions presented in Anthropocene discourse by means of graphs and algorithmic projections. The results of this are climate refugees, "abandoned villages, hopeless landscapes, [and] deserted beaches of these tiny towns and islands" (24; 175) rendered uninhabitable by a combination of industrial pollution and human violence between local militias and state military. These quotidian preoccupations of the novel's narrativity are what Louise Green in a relatively similar reading describes as "the insignificant details which are also the substance of the real" climate situation, but which often get "lost in the process of abstraction" (2018, 249)—the means by which the Anthropocene discourse articulates. The narrative strategy of infusing globalized climate discourse with localized substance of the real (actual violence) is not in any way another case of postcolonial *victim mongering*—far from it. Rather, Habila's text is an attempt, perhaps, to bring the Anthropocene debate back to the concrete substances of its implications by scripting out of the grand scales of its thoughts the particular ways in which to track and apprehend the extant dimensions of its present impact to better conceive of a future that befits the Anthropocene time scale.

Thus, Habila turns his writerly gaze on poor fishing communities and nondescript fisher-folks located along the Delta who, faced with terrible alternatives, must make a choice: either surrender their lands and familiar existence to oil extraction, and in the process benefit from its ephemeral gains, or resist the promised lucre, albeit shortly, until forcibly removed by the militarized forces that govern oil extraction, including the Nigerian postcolony. In the novel, communities that give in to the latter option get compensated in exchange for their land with flat screen TVs, DVD players, and second-hand four-by-fours. But these soon break down: "the cheap televisions and DVD players … gone … the rest of the money thrown away in Port Harcourt bar rooms, or on second wives and funeral parties, and now they are worse off than before" (Habila 2010, 29). There is a third category, though: the ones who neither actively resist nor blindly join in the consuming frenzy of the oil money, but are nevertheless caught up in the inevitable upheavals that ensue. Enter Boma, a character who emerges in the novel's end as the oil survivor and embodiment of hope in the post-oil world that the novel gestures to in the end.

To be sure, although the narrative appears woven around the search by two journalists, Zaq and Rufus, for the kidnapped wife of James Floode, a British oil company worker, it is also a journey into self-awareness by the protagonist, Rufus. It is a coming to terms with the fate of his family caught up—as part of the third category—in the social and ethical collapse of their community as consequence of oil extraction. Rufus's sister Boma is a victim of petroleum fire disaster caused by their father, who is himself a victim of circumstance, driven to illegal oil trade after losing his oil company job, "just like half the town. They all worked for the ABZ oil company, and now the town, once awash in oil money, watches in astonishment as the streets daily fill up with fleeing families, some returning to their hometown and villages" (61), others seeking refuge in nearby villages and islands in search of alternative means of livelihood.

The petroleum fire takes the lives of many in the village, including the father of Rufus's best friend. Even survivors are maimed permanently: families disintegrate, Boma's face is permanently scarred, their mother grieves inconsolably, and their father is arrested and jailed. As the novel ends, Boma comes to terms with her condition. She finds a home among a community of worshippers on Irikefe Island, a religious community that experiences its own trauma by the fact of its proximity to the unfolding violence and environmental catastrophe—their shrine desecrated, religious figurines destroyed, and some of its adherents killed or driven into the ocean. As Boma arrives in this community, her fate is soon tied to that of the worshippers, united by their experiences of trauma and of displacement as discounted victims of extraction. However, these villagers soon begin to rebuild from the rubble of their collective suffering: "Boma was with the group of women at the hearth. I could see her from here, her red blouse standing out in the cluster of white robes around her … she looked really happy, and for a moment I almost started to believe that the worst really was over" (166). The narrator (Rufus), who appears to witness this process of transformation, describes the scene by juxtaposing the moods of grief and resilience through concatenating images of blood and the colors of red, representing pain/affliction and experience, and white, representing purity and resilience: "Some tried to wash the blood spots off their white robes, without much success" (165). The image of bloodstained robes indexes the resilient transition to a new possibility beyond the one offered by oil. It shows how an attempt to begin anew need not, must not, be oblivious of the incumbent but, rather, must proceed from the experience of the past, must begin in the wake of the defunct regime—that is to say, not in spite of but because of the end of the current regime—and, therefore, has to carry vestiges of that experience.

Conclusion

My objective in this chapter has been to position oil landscapes in the Global South alongside the concerns that animate the Anthropocene discourse, focusing on recent representations in the Niger Delta as examples of a post-oil envisioning. The Nigerian artist and writer whose work I have discussed here are not oblivious to the prevailing debates that frame the Anthropocene; Zina Saro-Wiwa and Helon Habila are highly educated individuals at the top of their crafts, living and working in the United States. Yet, their texts are shown to present the reality (depicted in art forms) of climate change and the Anthropocene as a profound actuality of place happening in real time, whose effects are no longer the stuff of projections and possible obviations. Moreover, their vision of the Anthropocene, even though conceived within the framework of literary apocalypticism, is not by any stretch a dark one; rather, it is generative, pulling toward a post-oil possibility within which the Anthropocene might unravel, but not without accounting for the ways in which its current impacts *actually* bear out in local landscapes.

In other words, the texts of apocalyptic realism frame climate change as events, rather than abstract processes, putting environmental issues in graphic contexts rather than discursive abstractions. Here, then is a possibility yet to pique the interest of those experts and scientists tasked with framing a way out of the impasse that besets our blue planet. Habila and Zina, in their individual mediums, mobilize idiosyncratic performances, images, and metaphors to put abstract processes of the earth's changes in graphic contexts as events happening in concrete time, out of which equally concrete steps, albeit modest ones, can be taken to begin envisioning how to live with (and within) the new epoch of the Anthropocene.

Notes

1 See, inter alia, Steffen et al. 2011.
2 See, inter alia, Eghosa Osaghae 1995; Okome 2000; Obi and Rustad 2011; Ross 2012.
3 Ato Quayson has an excellent discussion of this notion in his book *Calibrations* (2003).

References

Aghoghovwia, Philip. 2017a. "Poetics of cartography: globalism and the 'oil enclave' in Ibiwari Ikiriko's *Oily Tears of the Delta*." *Social Dynamics* 43 (1): 32–45.

____. 2017b. "Nigeria." In *Fueling Culture: 101 Words for Energy and Environment*, edited by Imre Szeman, Jennifer Wenzel, and Patricia Yaeger, 238–41. New York: Fordham University Press.

Altvater, Elmar. 2016. "The Capitalocene, or, Geoengineering against Capitalism's Planetary Boundaries." In *Anthropocene or Capitalocene? Nature, history, and the Crisis of Capitalism*, edited by Jason Moore, 138–52. Oakland: Kairos.

Caminero-Santangelo, Byron. 2014. *Different Shades of Green: African Literature, Environmental Justice, and Political Ecology*. Charlottesville: University of Virginia Press.

Chakrabarty, Dipesh. 2012. "Postcolonial Studies and the Challenge of Climate Change." *New Literary History* 43 (1): 1–18.

____. 2013. "History on an Expanded Canvas: The Anthropocene's Invitation." Keynote speech presented at *The Anthropocene Project: An Opening,* Haus der Kulturen de Welt, Berlin, January 13. https://youtu.be/svgqLPFpaOg.

____. 2017. "Anthropocene 1." In *Fueling Culture: 101 Words for Energy and Environment*, edited by Imre Szeman, Jennifer Wenzel, and Patricia Yaeger, 39–42. New York: Fordham University Press.

Chomsky, Noam. 2018. "The Anthropocene, 6th Extinction, and Climate Change." Lecture presented at St Olaf College, USA, May 6. https://youtu.be/kjIsuGHdbnw.

Crist, Eileen. 2016. "On the Poverty of Our Nomenclature." In *Anthropocene or Capitalocene? Nature, History, and the Crisis of Capitalism*, edited by Jason Moore, 14–33. Oakland: Kairos.

Crutzen, Paul and Eugene Stoermer. 2000. "The Anthropocene." *IGPB (International Geosphere-Biosphere Programme). Newsletter* 41 (May): 17–18.

Crutzen, Paul. 2002. "Geology of Mankind: the Anthropocene." *Nature* 415: 22–23.

Crutzen, Paul and Will Steffen. 2003. "How Long Have We Been in the Anthropocene Era?" *Climate Change* 63 (3): 251–57.

Gelpke, Basil, and Ray McCormack, dir. 2006. *A Crude Awakening: The Oil Crash*. Lava Productions AG. DVD, 83 min.

Green, Louise. 2018. "Modernity's dirt: carbon emissions and the technique of life." *Social Dynamics* 44 (2): 239–256.

Habila, Helon. 2010. *Oil on Water*. London: Penguin.

Klein, Naomi. 2014. *This Changes Everything: Capitalism vs The Climate*. New York: Simon & Schuster.

Mbembe, Achille. 2000. "At the Edge of the World: Boundaries, Territoriality, and Sovereignty in Africa." *Public Culture* 12 (1): 259–84.

Mitchell, Timothy. 2009. "Carbon democracy." *Economy and Society* 38 (3): 399–432.

Moore, Jason W., ed. 2016. *Anthropocene or Capitalocene? Nature, History, and the Crisis of Capitalism*. Oakland: Kairos.

Morton, Timothy. 2013. *Hyperobjects: Philosophy and Ecology after the End of the World*. Minneapolis: University of Minnesota Press.

Nixon, Rob. 2017. "Anthropocene 2." In *Fueling Culture: 101 Words for Energy and Environment*, edited by Imre Szeman, Jennifer Wenzel, and Patricia Yaeger, 43–46. New York: Fordham University Press.

Obi, Cyril, and Siri Rustad, eds. 2011. *Oil and Insurgency in the Niger Delta*. Uppsala: Nordic Africa Institute.

Okome, Onookome, ed. 2000. *Before I Am Hanged: Ken Saro-Wiwa, Literature, Politics, and Dissent*. Trenton: Africa World Press.

Osaghae, Eghosa. 1995. "The Ogoni Uprising: Oil Politics, Minority Agitation and the Future of the Nigerian State." *African Affairs* 94: 325–44.

Quayson, Ato. 2003. *Calibrations: Reading for the Social*. Minneapolis: University of Minnesota Press.

Ross, Michael. 2012. *The Oil Curse: How Petroleum Wealth Shapes the Development of Nations*. Princeton: Princeton University Press.

Saro-Wiwa, Ken. 1995. *A Month and a Day: A Detention Dairy*. London: Penguin.

Saro-Wiwa, Zina. 2018. "Karikpo Pipeline (2015) 5-channel video installation 27mins 31secs." Zina Saro Wiwa (website). Accessed August 25, 2018. www.zinasarowiwa.com/video/karikpo/.

Sartre, Jean-Paul. 1961. Preface to *Wretched of the Earth*, by Frantz Fanon, 7–31. London: Penguin.

Steffen, Will. 2010. "The Anthropocene." Filmed November 4, 2010 in Canberra, Australia. TEDx video, 18: 15. https://youtu.be/ABZjlfhN0EQ.

Steffen, Will, Jacques Grinevald, Paul Crutzen, and John McNeil. 2011. "The Anthropocene: conceptual and historical perspectives." *Philosophical Transactions of the Royal Society A: Mathematical, Physical and Engineering Sciences* 369 (1938): 842–867.

Steffen, Will. 2013. "The Anthropocene: Where On Earth Are We Going?" Keynote speech presented at *The Anthropocene Project: An Opening*, Haus der Kulturen de Welt, Berlin, January 10, 2013. https://youtu.be/T8U6y4UNXRQ.

Zalasiewicz, Jan, Mark Williams, Will Steffen, and Paul Crutzen. 2010 "The New world of Anthropocene." *Environmental Science & Technology* 44 (7): 2228–2231.

3

FIRE, WATER, MOON

Supplemental seasons in a time without season

Anne-Lise François

The following essay is to be read as a companion piece to "Ungiving Time: Reading Lyric by the Light of the Anthropocene," published in the volume edited by Tobias Menely and Jesse Oak Taylor *Anthropocene Reading: Literary History in Geologic Times* (2017). In that essay I draw on Andreas Malm's conceptualization of the Anthropocene as a change in the Earth's relation to the Sun that arises when fossil fuel economies draw on and release, only to trap again, the stored energies of solar years past. In this essay I ask what has not changed in the Earth's relation to the Moon and consider the persistence of the multiple, secondary seasons that, determined by the Moon's monthly rotation around the Earth, remain embedded, and continue to make epicycles, within the Earth's single annual rotation around the Sun.

"A time out of time, in more senses than one"—is how Dana Luciano (2015) pithily refers to the post-Holocene, in a phrase that captures both the political urgency of having run out of time to avoid the worst impacts of climate change, and the existential crisis of having been ejected from the (more) normal rhythms of diurnal, circadian and seasonal time. In the nearly two decades since Paul Crutzen and Eugene Stoermer proposed the term "Anthropocene" to designate a new geological epoch defined by the impacts of human activity on the Earth's living system, numerous critics have rightly faulted the term for failing to discriminate between types of human activity, for discounting the long history of other kinds of mutual modifications between humans and their habitats, and for inventing a culprit—the "human species"—for what are in fact historically specific forms of violence: the historically determinate concentrations of power necessary to plantation-based modes of production, extraction, and exploitation.[1] Hence the proliferation of alternate terms such as the Capitalocene or the Plantationocene and of alternate dates to the one proposed by Crutzen and Stoermer—1784, the year of James Watt's invention of the steam engine—for the start of this new era. The date proposed by the climate scientists Simon L. Lewis and Mark A. Maslin (2014) is 1610 because it marks both the lowest level in the drop in atmospheric carbon dioxides attributable to the reforestation that followed the deaths of 50 million Amerindians post-European conquest and also the point of no return from the homogenization of planetary biota initiated by the same colonial contact. As Luciano (2015) argues, by rendering the stratigraphic record interpretable as evidence of genocide—as the inscription of mass murder—Lewis and Maslin enroll climate science among the "numerous cosmologies [that] hold that the Earth will remember acts of intra-human violence, that the planet itself will testify to the brutality humans have inflicted upon members of their own species." I will return in the following sections to this figure of the stratigraphic record as the encryption of human history, for the drop in CO_2 levels represents the most minimal, anti-representational of epitaphs—a dip in the line—that encrypts in the precise sense of marking, without necessarily making legible (or grievable), the disappearance of human societies.[2]

Early twenty-first century work on the Anthropocene can thus be understood as a belated and uneasy recognition of the Earth's power to assume a task which European philosophy might once have assigned exclusively to culture: that of archiving human history (by becoming the storage house for non-erasable anthropogenic deposits.[3] So, in *The Shock of the Anthropocene*, Bonneuil and Fressoz (2015, 8, 13) include among destructive anthropogenic effects, in addition to the warming of the Earth's temperature caused by the burning of fossil fuels and to the loss of biodiversity caused by large-scale industrial agriculture and urbanization: the "modification of the continental water cycle" (through the draining of wetlands, damming, and irrigation), and the "deposit in the planet's ecosystems over the last 150 years" of "entirely new substances (synthetic organic chemistry, hydrocarbons, plastics, some of which form a new type of rock, 24 endocrine disturbants, pesticides, radionuclides dispersed by nuclear tests, fluoride gases)." Not without irony, this renewed attention to the Earth's capacity for monumentalization and encryption follows on the heel of analyses of late capitalism focusing on the dematerialization of finance and labor as an intensification of contingency, mobility, groundlessness, temporariness, and liquidity.

As Bonneuil and Fressoz argue (32–33), far from a kind of mastery, the supposed elevation of "man" or "the human" to the status of geological actor capable of changing the face of the planet in fact spells the end to illusions of freedom from nature and definitively brings to a close a certain narrative of modernity according to which nature simply is what it is, while the human is defined as the (implicitly white, male) subject who gives itself a history through its transformation of the material world. The contradictions attending this simultaneous neutralization and activation of transformative power come into sharper focus when we consider that the notion of a signature so heavy that its erasure cannot be imagined perverts the nature of signing, or forces a change in what we mean by it (at least from a Derridean perspective), and does so to the same degree that the appropriation of lightness and temporariness as the prerogative of power perverts the nature of transience. Compare, for example, Bonneuil and Fressoz's (13) claim that such deposits "constitute a typical signature of the Anthropocene in the sediments and fossils in the course of formation" to Zygmunt Bauman's (2000, 13) claim in *Liquid Modernity*: "Travelling light, rather than holding tightly to things deemed attractive for their reliability and solidity—that is, for their heavy weight, substantiality and unyielding power of resistance—is now the asset of power. … Holding to the ground is not that important if the ground can be reached and abandoned at whim, in a short time or in no time." Both the figure of the perdurability of traces and the fantasy of their easy undo-ability read as twinned perversions of a lost (and perhaps never in fact enjoyed) capacity both to disappear and remain as the wind does.[4]

Heavy liquidity, sunken (unfloatable, non-buoyant) latency, immutable (non-malleable) disposability—with such figures we can begin to think together both the enormity of material consequences and their continued defiance of a conventional sort of realism, whereby only the actual or empirically present counts as present.[5] In "History of the Future", Barbara Adam (2010, 373) critiques a certain crudely rationalist way of conceiving of the future as an "aspect of mind, belonging to the realm not of the real but the ideal," and of thinking of the "not yet" as "not material, not factual and [as having] no reality status" because "only materialised processes of the present and past have the status of factual reality." According to Adam:

> in ancient times this would have been considered utter nonsense. With the rise of science, however, it became common sense … In our technologically driven world, where time-space-matter-distantiated outcomes may be latent for hundreds, even thousands, of years, this assumption is once more becoming untenable. In such a context we have to accept that the latent world of processes is real even if it is not tangible or material in the conventional sense of materiality.
>
> *(373)*

If capitalist modernity spins a story of ever-widening horizons of expectations, the Anthropocene is the name given within this narrative—often by those who never until now stopped to doubt its promises—for the blockage that results ironically from the attempt to appropriate the future as an "abstract, empty and quantifiable entity available for free unrestricted use and exploitation" (365).

This figure of a future blocked in advance by radioactive and carbon half-lives is, in turn, only the temporal inversion of the release into the present of the deep geological past represented by coal mining and the burning of fossil fuels. At least this is how in *Fossil Capital* Andreas Malm (2016) helps us conceptualize the Anthropocene—as an ontological break in and messing up of the Earth's relation to the Sun—when he draws on E.A. Wrigley's differentiation between "organic" and "inorganic" energy economies in terms of their relative dependence on or independence from the "yield of present photosynthesis" (20): where earlier agrarian societies still stood more or less in relation to the annual supply provided by one rotation around the Sun, the "present" of coal and post-coal economies expands to include the "stores of *past* photosynthesis" (21). Whether dated to the

Great Acceleration or to Columbian Exchange or to the Neolithic revolution, the Anthropocene is a figure for the *fast* consumption of deep time—a name for the sudden ability to access previously locked biotic and cultural abundance and the simultaneous power to *smoke* or destroy it forever. The Anthropocene, so understood, would be the name for the time (epoch) of the time (duration) of a final cigarette, each drag of which irrevocably consumes in a matter of minutes millennia of densely accumulated and long-deposited biotic information and cultural know-how. The figure of smoking deep time may be helpful in loosening the grip of the fixation on the consumption of fossil fuels per se. Plastic and electronic waste; the extraction of uranium and its conversion into nuclear energy, arms, and waste; the drying up within a mere 50 years of the American Southwest's millennia-old water tables following the invention of the centrifugal pump; and the prodigal squandering of soil fertility that results from industrial monocultures—each shares the same tripartite temporal structure: a present sandwiched between a past which it can only conceive as its prehistory—as the deposits which, accumulated from time immemorial, it consumes as if immediately available—and a future which it can only conceive as aftermath, as the interminable afterlife of what it will have left behind.

One obvious and direct consequence of the break with seasonal solar time is the dystopian logic of 24/7 decried by Jonathan Crary (2013) in *24/7: Late Capitalism and the End of Sleep*. The post-seasonal future of which climate scientists are warning is already here, to the degree that contemporary anthropogenic climate change originates in an economic system built on the homogenization of space and time, on the erasure of seasonal variation, and on the fiction of the permanent availability of labor, goods, and services. For Crary, the logic of 24/7 determining the round-the-clock illumination of homes, hospitals, factories, and solitary detention cells alike represents an assault on the right to sleep and on the capacity to absent oneself in periods of inoperative and non-serviceable darkness. But where does this logic begin? In the extension of daylight used in industrial poultry farms to force hens to lay eggs twice rather than once a day? Or does it lie already latent in the contingent moment in evolutionary history when humans along with other primates acquire the monthly menstrual cycle, a tool or temporal prosthesis freeing sexual reproduction from a specific season in the earth's solar year?

In Simians, Cyborgs, and Women, Donna Haraway writes of how early twentieth century primatologists translated their discovery that "[p]rimates possess the menstrual cycle" into the "fantasy-inspiring fact of constant female receptivity" (1991, 22). Should we then imagine the history of a certain kind of modernity as nothing more than the progressive decoupling of reproduction from the Earth's position with respect to the Sun, and the consequent intensification of forms of uninterrupted availability? An immediate answer would be "no," since the menstrual cycle does not in fact translate to women's permanent readiness for conception, and humans are further differentiated from other primates in "combin[ing] continual receptivity with loss of estrus" (Shaw and Darling 1985, 79). If seasonal or patterned time is time divided into strongly differentiated periods of "sleep" and "waking," activity and dormancy, receptivity and closure, then—to reread the accounts of evolutionary biologists—primate and then specifically human reproductive sexuality appears as the result of two decouplings from obviously or strongly marked patterned time, one nested within another: (1) procreation can occur every month rather than annually or biannually; (2) within this monthly cycle, the time of ovulation is not signaled as such and sex can happen and can also be refused at any time. Rather than translating to more reproductive sex or guaranteeing (greater) "female availability," concealed or "unadvertised" ovulation spells the inability to discern procreative sex from nonprocreative, the indistinction of fertility from mere receptivity. Intermittency is concealed and to the undiscerning (male) eye appears as uniformity, making it appear that x can happen at any time, without it being possible to determine the readiness for x. Misogyny may be the tale of the skeptical crises that ensue when women are no longer conscripted to sex by markers legible on their bodies at discrete periodic intervals, and instead retain a constant low-level capacity both to resist and to accommodate potential sexual partners.[6] The operative difference generating the illusion of surplus or permanent supply is in the order of potential rather than actual value, something that will be true in all of the examples that follow.

This much I argue in "Ungiving Time," an essay in which I ask about what, if anything, changes for human temporality when societies begin availing themselves of millennia of past solar years. In these pages, I turn instead to the Moon and to the continued help that might be proffered by the various, relatively unchanged lunar cycles. Amidst the anxieties provoked by both the prospect and reality of post-seasonality, it helps to remember that the relatively stable patternings of distributed abundance and dearth, scarcity and fertility we knew as the seasons were never simple, and certainly never simply distinct from one another.[7] So, in *A Midsummer Night's Dream* Titania refers to the "summer's middle spring"—a figure that might well send us into ecstasy as it did John Keats because of the way it swells with joyous self-reflexivity to trope on "spring" as something more than the calendar season—a figure for the latency by which one season may bear and give birth to another buried within it, just as the new Moon is said to hold the old Moon in her arms.

One of these micro-lunar seasons giving us a little winter and summer within each month is the cycle determined by the tilt of the Moon's path around the Earth with respect to the ecliptic. Crossing the plane of the Earth's orbit at a mean angle of a little over five degrees, the Moon ascends from Southern Celestial Hemisphere to Northern Celestial Hemisphere, rising a little higher in the sky each night and then descends from Northern Sky to Southern, rising a little lower each night, completing this cycle within on average 27 days, while the time between new Moons is closer to 29 days, as the Moon has to catch up with the Earth's own advance around the Sun. Determining whether the Moon is ascending or descending requires serial and relatively sedentary observation, which may explain the disappearance of the draconic month from popular consciousness: to determine whether the Moon is "riding high" or "low" (to borrow the metaphors of old farmers' almanacs), one has to observe the Moon with respect to the same fixed point, each night some

50 minutes later (because the Earth is also traveling), over a series of nights. Visualizing the higher or lower arc drawn by the ascending or descending Moon may recall the perhaps forgotten elementary or secondary school lessons about how the Sun appears to ride higher in the sky in summer and lower in winter. People who garden according to the Moon are in the habit of describing the three intercalated rhythms of ascent and descent—the Sun's movement across the sky in the course of the day, the rising and descending arcs it traces in the course of a year, and the Moon's in the course of the draconic month—as a kind of breathing, a pattern of release and contraction, an exhalation and expansion followed by a drawing close or drawing down of energies. Thus, morning is the spring of the day, evening its fall. Ascending arcs represent exhalations, descending arcs inhalations, something that will seem counter-intuitive until you imagine yourself as a plant, with your nervous system below the ground.

I offer this brief excursus into intercalated yearly and monthly rhythms as a way of qualifying my previous suggestion that we consider menses as prototypical of the independence from seasonal time that will extend the hours of (re)productivity late into the nights of industrial capitalism. The suggestion may be useful for illuminating a certain conceptual continuity in the fiction of permanent availability, but it should not be taken as dismissing the relations—however open and indeterminate—between lunar cycles and cycles of ovulation, nor as discounting the many different kinds of reproductive seasons punctuating earthly cycles. In the remaining pages, I would like to turn to accounts that recover or bear witness to the threading of the seasons into ever more discrete, determinate, and delicately intercalated units of time. My examples are deliberately chosen from radically different traditions, with little in common except their foreignness to modern clock time and their ability to serve as counterexamples to the catachronic relation to the Sun resulting from the ability to access its accumulated energies and store them again as potential, which goes by the name of the Anthropocene. While only the last example is strictly literary, and only it addresses the relation between poetic rhythm and lunar time, all three examples reflect the cultivation of phenological patterns in precolonial and pre-enclosure modes of knowing.[8] The reader impatient for the Moon and for the question of lunar seasonality will have to await this last example, the other two being only conceptually related in their attention to complex but relatively regular modulations in patterns of waxing and waning energy.

My first example challenges the zero-sum game, according to which the relative dominance of plants can only indicate the relative absence of humans and vice versa, implied when Lewis and Maslin interpret the stratigraphic drop in CO_2 levels as indicative of the reforestation that followed the genocide of Amerindians post-European conquest. The evidence offered in M. Kat Anderson's (2005) *Tending the Wild: Native American Knowledge and the Management of California's Resources* to some extent confirms Lewis and Maslin's hypothesis—prairies and meadows did give way to denser forests following the killing and forced removal of California Indians—but attention to Anderson's argument would also remind us that indigenous peoples were not simply swallowed by the forests that took their place; their relation to the ecological climax community was one of continued and reiterated delay and deferral, rather than simple opposition and negation. According to Anderson, the extraordinary diversity of, and micro-differences found among, California's wildflowers can be read as an expression and persistence of disappeared California Indian practices even if these practices themselves appear to have been (for now) violently discontinued:[9]

> Many of the biologically richest of California's habitats were not climax communities at the time Euro-Americans arrived but instead were a rich mosaic of ecological succession, or fire sub-climaxes, intensified and perpetuated by seasonally scheduled burning. … Heightened species diversity, abundance, and density have been associated with regular, intermediate-intensity, spatially heterogeneous disturbance. Based on this relationship, it can be hypothesized that the disturbance by California Indians' use of fire in a variety of ecosystems, occurring at intermediate intensities and frequencies, promoted a maximally heterogeneous mosaic of vegetation types and increased species diversity.
>
> *(156, 153)*

Anderson invites us to read the work of California Indians in the pockets of wildflowers that remain like outcroppings in the absence of the regularly timed disturbances that allowed them to take hold in the first place: such flowers may be living beings but they are also graveyards, museums, ruins, traces, forensic evidence of mass murder as well as auto-representations of the pollinating practices that dilated time and delayed the onset of climax communities. Discernible in the evidence through which Anderson sifts, in the oral narratives of survivors as well in the notes of European naturalists, is a formalization of time comparable to that of shepherds' calendars or farmers' almanacs. The sense that emerges is of the organization of time into distinct intervals of "on" and "off"—overlapping cycles that extend sometimes to two years, or three, or five; "seasons" not partitioned according to the earth's position relative to the Sun nor concluded within a single rotation around the Sun, yet still in some kind of determinate relation to the work of photosynthesis performed within that period. If I quote at length from Anderson's account of the different regulations determining the frequency and timing of different harvests, it is because only multiple excerpts can do justice to the elegiac work she is performing in allowing into her prose some sense of the extensive, circular and repeatable time that would have been necessary for these now mostly inoperative calendars to be established:

> Much of the rich material disclosing the ancient management of wilderness lies in the dusty diaries and handwritten notes of anthropologists and the eyewitness accounts of early European settlers. For example, Kroeber's 1939 field notes, housed at the Bancroft Library at the University of California, Berkeley, record that the Yurok of northwestern California practiced burning

> at a frequency that was appropriate for each cultural purpose: burning of hazelnut for basketry occurred every two years; burning under the tan oaks to keep the brush down took place every three years; burning for elk feed occurred every fourth or fifth year; burning in the redwoods for brush and downed fuel control occurred every three to five years. These observations did not change his thinking about "hunter-gatherers," nor did he publish them.
>
> *(126)*

> Hidden in the simple act of gathering lay sophisticated rules that safeguarded the plant stock from overharvesting. An overarching gathering rule was, Spare plants or plant parts; do not harvest everything. … There were many more specific rules in California, such as, Take roots from one side of the tree only. When digging ponderosa pine roots for baskets, the Maidu cut roots from one side of the tree one year and from the opposite side the next year, so as not to kill the tree. …
>
> *(128)*

> Seed beating may have enhanced the herb population in a number of ways. In addition to scattering seeds around the collective area, the method ensured that only ripe seeds were collected, leaving immature seeds to develop fully and replenish the stock. Breaking off fruit heads or uprooting whole plants and removing them from the site, in contrast, easily could have extirpated wild populations of annual plants in a matter of years or decades. Leaving most of the plant behind was important in terms of lightening the load a woman carried in her burden basket. But it also ensured that the biomass would eventually be recycled in that place. This fact becomes important in the application of another management practice—the periodic setting of fires. The seed-bearing grasses and wildflowers, as they dry and wither, become the fine fuels that will carry frequent, light surface fires.
>
> *(129)*

To "manage" here means to divert and fray what we now more or less ignorantly refer to as the single annual "fire season" by diversifying it into multiple seasons for different kinds of fire. In remembering that the frequent setting of such low-level fires would have prevented the kind of far more lethal top-crown forest fires so destructive to California's forests in recent years, there is cause for grief in the way national forest policy continues to misrecognize this ecology, but also a chance to recover a "seasonality" that was less about living in accord with set periods of solar proximity and distance than of alternatively accentuating, minimizing and pluralizing these fluctuations, disturbing and distributing their energies across time and thereby deferring an otherwise violently destructive punctual event. In the same way, in protracting and extending the open spaces for wildflowers, thereby deferring the onset of what twentieth-century ecologists would call the forest climax community, the seed-gathering and -disseminating practices of California Indians compose a present that only appears unchanging, stable or singular: a threading together of a series of successive prolongations of transient meadows.

The figure of an illusory inertia or precarious durability, composed of relatively small or short-lived extensions, whose shortness means they must constantly be renewed, is the link to my next example drawn from an entirely different context: Anupam Mishra's (2000) explanation of how for hundreds of years traditional water-collecting methods in the arid Indian state of Rajasthan were able to make the water from the annual meager rainfall last from one season to the next, sustaining significant populations without irrigation or dependence on water importation from other regions.[10] Mishra explains how *kuîns*, little household wells, neither deep nor wide, which reach only to the area where water gets trapped between benches of gypsum and surface sand, make available again rain that has already fallen (*kuîn* is the feminine diminutive of the masculine *kuân* for well):

> a well is supposed to capture groundwater, but this not how a kuîn collects water. What it collects is rain water, and in a special way, even when it hasn't rained. For what is found in the *kuîn* is neither water that runs on the surface nor the stagnant (dormant) water below ground. It's a complicated affair that derives from the *neti* ... *neti*, the "neither ... nor" of Sanskrit that allows the cosmic principle to be described by negation in the Upanishads, for example.
>
> *(70, my translation)*

The miracle occurs thanks to the relative shallowness and narrowness of the *kuîn* as drops of water sweat or seep out very slowly from the humidity lodged in the sand; they would evaporate if the opening were any wider. Where gypsum is present to prevent rainfall from mixing with the brackish groundwater below, "all the drops that fall amass themselves in the sand and change into humidity. If a *kuîn* is built there, its stomach, the empty cavity, retransforms this accumulated humidity into drops" (72). Mishra attends to the specificity of linguistic terms used to capture the distinctness of this capillary water that, neither rainwater nor groundwater, lives in the sand: it is "measured not in inches or centimeters but in *rejâ*, *rejo*, a term that signifies 'fragment,' and *rejâ* doesn't measure rain that has fallen on the earth's surface but rain that has accumulated underneath its surface" (73).[11]

The salient detail for me is that in one day a *kuîn* is barely able to collect enough water to fill two or three pitchers of 20 liters; there may be enough but no surplus, nothing left over to lay away as security or trade as capital. We have here then a form of moment-to-moment (or hand-to-mouth) storage and unreservable accumulation—a method of protracting the otherwise fallen, ephemeral and passing—that remains antithetical to the standing reserves, or stockpiles of stored energy awaiting use, of petrocapitalism. To the latter, such a system can only appear poor because on any given day there is only enough water to get through to the next. In this sense, what Mishra calls the system of relays and "radical decentralization of the work of water," by which "is disseminated among everyone the responsibility of every drop," is temporal as well as spatial: here, as with the forms of capturing wind and water energy analyzed by Malm, to use is to spend (more or less everything) as you go and as it passes; because only time does the work of making something available, there is no illusion, as with coal or its successors, of abundance in the present or richness out of time. By the same token, there is no incentive to save (no pressure not to share or squander) what will in any case be supplied again on the morrow.

The overlap in figurative and conceptual terms—collection, concentration, captivation, etc.—between Mishra's (2000) discussion of traditional water-collecting methods and Malm's (2016, 38–39) account of wind and water prior to the development of coal-based capitalism, as "sources of energy originating in the sun that flowed through the biosphere, uncaptured by photosynthesis, accessible for direct collection and concentration by prime movers designed for the purpose," may serve as a salutary reminder that not all forms of collection or storage are synonymous with violent expropriation and that the problem is not with collection or storage or temporal prosthesis per se. In rejecting Wrigley's Ricardian-Malthusian account of the advent of coal as a chance to break free of "the bounds of the organic economy" and leap "into continuous growth," Malm (2016, 20–25) at one point quips that the water mills that preceded coal as Britain's primary source of energy were not plants requiring to be fed more Sun than was available in any given year. The image inadvertently brings out the oddity of conventional agriculture and the violence necessary to its annual forcing of the reproductive parts of grain crops. Perpetually delaying the onset of other vegetation while harvesting plants at the point of reproductive maturity, freezing the possibility of reproduction while pressing pause on the cycle before it can occur, the cultivation of annuals would seem, from one point of view, only to intensify, accelerate, and replicate at a much larger scale the distention of ecological time of California Indian gardening practices. At what point does this difference in degree become a difference in kind? In "Ungiving Time," I suggest that Wrigley's historical argument about a shift from dependence on, and movement with, organic energy cycles to a more perpendicular inorganic energy extraction could easily be relativized by remembering that agriculture is often also credited with instituting a shift from living within the constraints of the moment to tapping into deep time. Earlier I sought to mark the irony that this so-called release from the near present should have resulted in the kind of entrapment in permanent immediacy decried by Crary, and made obligatory the kind of "realism" that consists of taking precautions to ensure that everything be potentially accessible at all times.[12] Rather than lay the blame for the Anthropocene at the feet of agriculture, as if within it there were only way of extending and storing an otherwise passing solar energy, these disparate examples suggest the need to examine different modalities and temporalities of collection and release, some of which are more susceptible to the captivation of surplus as marketable reserves necessary for speculative capitalism.

I began by noting the degree to which Anthropocene discourse focuses on humanity's power to expand and compress, accelerate or distort circadian rhythms in terms of the *Sun* and solar years. I'd like to turn now to the provocation with which I began this essay: the thought that in the midst of the extreme climate instability unleashed by fossil capitalism's extraction and commodification of past stores of solar energy, the cycles associated with the Moon as it journeys around the Earth, and with the Earth through the constellations, have remained relatively stable. I hope readers will forgive me for what will doubtless feel like another leap across centuries and cultural contexts. One ironic measure of the specific untimeliness of the Anthropocene (specific because there are, of course, other kinds of aseasonality and ways of being out of synch with planetary movements) would be the early twenty-first-century reception of Sappho's "midnight fragment," verses which rely on their recipient's ability to tell time by the position of the Moon and stars:[13]

> δέδυκε μὲν ἁ σελάννα
> καὶ πληΐαδες' μέςαι δὲ
> νυκτες, παρὰ δ' ἔρχετ' ὤρα
> ἔγω δὲ μόνα κατεύδω
>
> Moon has set
> and Pleiades: middle
> night, the hour goes by,
> alone I lie.
>
> *(Carson 343)*[14]

In *Figures of the Thinkable*, Cornelius Castoriadis (2007) castigates "contemporary readers" who remain oblivious, he claims, to all that these four short lines impart by way of empirical information—a gift they make without giving away that they are doing so:

> It's the middle of the night and the moon has already set. This is meaningless for our contemporaries ... They have no realization that because the moon has gone down before midnight, it is somewhere between the beginning of the new moon and the first quarter, and therefore the beginning of the lunar month (the measure of time for all ancient peoples). But the Pleiades too have already set. ... It is springtime—and actually early springtime—for that is when the Pleiades set before midnight. As the year grows older, the later and later they set. Sappho has set, and the *ōra* passes.
>
> *(23, translation modified)*

While Castoriadis is correct that in Greece the Pleiades set before midnight between mid-January and late March, his wording curiously anticipates the winter that is past as to come. For, in fact, like all stars, they rise and set about four minutes earlier every day, and with the advance of spring they would be moving toward the time of their non-visibility in the night sky and toward their heliacal rising in early May: the earliest point when, after rising with and being obscured by the Sun, they rise just early enough to be briefly visible before dawn.[15] For Hesiod this moment of reemergence from out of the Sun's shade signals the start of the harvesting season: "When the Pleiades, daughters of Atlas, are rising, begin your harvest, and your ploughing when they are going to set" (1914, 385). Anyone who took this to be a reference to diurnal rising and setting would be at a loss to know when *not* to sow or harvest, but it is doubtful Hesiod's contemporaries would have made such a mistake. Indeed the unusualness of a reference to the Pleiades' nightly, rather than yearly, setting prompts Paula Reiner and David Kovacs (1993, 148–149) to suggest the text of Fragment 168b is corrupt and to propose emending the line to "the Pleiades sail aloft," taking their cue from A.E. Housman's "The weeping Pleiads wester,/And the moon is under seas." Yet I would argue that in supplying a second verb, Housman is not so much correcting what Reiner and Kovacs perceive as flat, redundant, boringly finished "perfective states" (154), but responding to a complex if submerged drama of anticipation and regret, convergence and disaggregation defined by partial synchronies where arrival and parting occur across the slightest of time lags. The time of the Pleiades' keeping company with the night sky is over. Meanwhile the Earth's nights are shortening with theirs. But even as the Sun rises earlier and earlier, there will come a point when it will be as though there was less sunlight than on the previous day because at their heliacal rising the Sun will have retreated with respect to the Pleiades, providing in the midst of returning light a measure of scarcity—a hair's breadth gap within the hour of ripeness.

All of this may be latent within the senses of senses of ὤεν = *[h]ōra* as season and opportune moment to which modern readers, conscious only of clock time, remain oblivious according to Castoriadis:

> But *hōra* is also the time when a thing has its hour, when a thing is truly fine and lovely, and therefore, for human beings, the height of youth. In the *Banquet*, Alcibiades tells how he tried to sleep with Socrates, but arose in the morning having been no more molested (*katadedarthēkōs*) than if he had slept with his father or his brother. He concludes: Socrates is a *hubristēs*, a man who insults others, to the point where he scorned (*katephronēsen*) my *hōra*, he spurned my youth, my beauty, the fact that I was ripe for the picking like a beautiful erotic fruit.
>
> *(23)*

The comparison beautifully if ironically brings out the latent eroticism of Sappho's fragment, for unlike Alcibiades, Sappho doesn't vociferate: the last line may imply complaint toward an awaited lover who it is by now clear is not coming, but the "alone" could just as well be relative to the celestial bodies with whom she has kept company and whom she will soon follow in sequence—if she has not already. At least this is how Castoriadis's precise wording—"the *ōra* passes and Sappho has set" interprets the verb κατεύδω, which can mean "lie [with]" or "lie down," or "sleep." Remembering the utopian senses with which Crary invests the ability to sleep, to surrender, to give in to rest, it is impossible to say whether the "ego" lies awake while the rest of the world takes its course, or has met its right end, for the sense of ripe over-fullness or erotic readiness, in any case, makes one indolent and hence susceptible to the hubristēs of letting the hour pass and squandering it alone. Castoriadis's gloss on the Greek is helpful here again:

> *Deduke*, from the verb *duō,* means "has plunged." In Greece,—with its two hundred inhabited islands and some ten thousands kilometers of coastline,— the sun, the moon, and the stars do not go down, they plunge.
>
> *(21–22)*

The gap between the active, once-and-for-all, irreversible commitment of "plunge" and the ongoing steadiness (patience?) of "lie" leaves open the question of whether she is still waiting or also passing.

As Clay and Castoriadis note, this question as to whether the "ego" is doing something *with* or in distinction (and isolation) from the other night-wanderers is also a function of the ambiguous particles μὲν. … δέ, which can be translated as "one the one hand … and on the other," but carry the same openness of meaning as the English "while"—perhaps the most important word in the study of phenological relations under climate change: does simultaneity of occurrence indicate necessary conjunction, happy coincidence, or an adversative "even so"? The uncertainty of the (for now) stationary self's relation to the other moving bodies with which it temporarily coincides is perhaps also embedded within the phrase παρὰ δ' ἔρχετ' ὤρα: but the hour passes (me?) by; the hour comes as it goes (what arrives disappears), and the self may be nothing but a name for what gives directionality to the otherwise ambiguous motion of the verb *erkhomai* (to come/ to go), just as the counter-turns are a means of cutting this otherwise ceaseless flow (compare the slow agony of André Bonnard's "l'heure s'égoutte" [the hour drains away, drop by drop] to Argyro Loukaki's Heraclitean "Midnight comes—and goes, the hours fly").[16]

In any case, according to Castoriadis, it is the poem that now lies alone, bereft of readers capable of reading it. Yet, if we are not in time for the poem, if the moment of its easy legibility has passed—the moment when it could count on its recipient to know it was about setting early (or not) as the Moon does in its spring, and as the Pleiades do in the spring of the year—but if we still take it to be the record of a fundamental mismatch between the time of human desire and cosmic time, then, in another sense, this means the poem's time has finally arrived—a time ripe for its missing as it rises up to us from the sea of drowned knowledges, spelling out the memory of a lost literacy, and making legible an inattention so gross the celestial bodies might as well have definitively set without return.

In turning away in these pages from the heliotropism of Anthropocene discourse to the other, more shadowy, secondary, if no less governing, measure of lunar time,

I have been taking my cue from Sappho's famous transposition of the Homeric epithet "rosy-fingered" from Dawn (Αὔως or Eos) to σελάννα in Fragment 96:

> But now she is conspicuous among Lydian women
> as sometimes at sunset
> the rosyfingered moon (βροδοδάκτυλος σελάννα)
> surpasses all the stars. And her light
> stretches over salt sea
> equally and flowerdeep fields.
> And the beautiful dew is poured out
> and roses bloom and frail
> chervil and flowering sweetclover
> (Carson 2003, 191)

The lines borrow from Homer's epithet the memory of tactility within a light so light that it only grazes or "brushes," without burning, in order to recast this touch as the coolness with which the "rosyfingered moon" evenly distributes her light across salt sea and sweet meadow, as if drawing the salt from the sea to draw flowers out of the earth.[17] Here is how Page DuBois articulates the latent feminism of this turn:

> I earlier argued that it seems to represent a celebration of the unplowed earth, the terrain spontaneously yielding up flowers rather than the life-sustaining grains of cultivated fields. In contrast to discourses that honor the earth as lit up by Homer's "rosy-fingered Dawn," that make an analogy between the plowed furrows in the earth and women's bodies, and that see agriculture as a crucial and sustaining practice of life in the ancient city, Sappho seems to privilege the female moon (*a brododaktulos mēna*) and feminine dew (*a d'eersa kala*), and the flowers that bloom at night.
>
> *(191)*

But in reassigning the epithet "rosy-fingered" from dawn to the Moon, Sappho doesn't just transpose day to night, borrowing from Homer as the Moon does the Sun, to convert and transmute a solar energy into a lunar one; she also retains the sense of spring-like earliness and renewal, for a Moon that would be visible at sunset would be somewhere in between a new, waxing (dawning) Moon and a full Moon. In this sense, the feminine shadowy penumbra is not just, as DuBois suggests, an alternative to the Sun-driven world of masculine extraction of female energies, but a repetition and embedding within it of the Sun's cycles of growth, except at a smaller scale and in pluralized form. DuBois herself goes on to cite Anne Burnett's reminder that "according to ancient belief, it was the moon that brought plants to their perfection, and it was by moonlight that flowers were gathered for the beneficent magic and for medicines" (308).

Burnett's reference is in turn to A Delatte's *Herbarius: Recherche sur le cérémonial usité chez les anciens pour la cueillette des simples et des plantes magiques*. The ancient belief cited by Burnett and Delatte concurs with the present practice of those who sow and harvest different kinds of crops according to the lunar phases, because they take seriously the idea—already partially expressed in Sappho's lines—that the Moon exerts an influence on earthly grasses comparable (if weaker) to what it does on the oceans', and recognize that this pull varies according to the Moon's position in its cycle.[18] I am suggesting that it is far better to be able to trace a genealogy of attention to such responsiveness back to Sappho than to have to see in these farming practices only a reactionary expression to early twentieth-century methods of industrial farming.[19] In my wanderings on small-scale organic farms, I have met people who practice farming according to the lunar calendar, not from scientific conviction per se, but as a way of organizing an otherwise unmanageable number of tasks—each onto the day, or night, as it were. Such farming does not want to colonize the night, as DuBois claims conventional agriculture does the day, but wants, from convenience and perhaps also from principle, to make itself pastoral, to work in relation to already ongoing cycles, to avail itself of the powers of dormancy, and to let other, otherwise independent, temporal processes do a lot of the work, or lend a shaping hand. Over against the catastrophic tenor of so much Anthropocene discourse that, in the manner of settler colonialism, rushes to declare everything over so to make a tabula rasa for development, it behooves us to remember, along with the Moon, the kind of ongoing knowledge practices of extension, protraction, collection, and diversification discussed in this essay.[20]

Notes

1 The critiques have not ceased to multiply; see also, in addition to Bonneuil and Fressoz, Donna Haraway, *Staying with the Trouble: Making Kin in the Chthulucene* (Durham, NC: Duke University Press, 2016) and Jason Moore, *Capitalism in the Web of Life: Ecology and the Accumulation of Capital* (New York: Verso, 2015).

2 For a moving discussion of the latent elegiac potential within these stratigraphic maps, see also Vincent Bruyère's "Ice Cold Grief: 1610 and the Orbis Hypothesis."

3 For a critique of the ways in which the renewed recognition of the Earth's agentive capacities to remember, on the part of Bruno Latour and others, ignores and erases Indigenous knowledge-paradigms, see Zoe Todd, "An Indigenous Feminist's Take On The Ontological Turn: 'Ontology' Is Just Another Word For Colonialism," *Journal of Historical Sociology* 29: 1 (March 2016) https://doi.org/10.1111/johs.12124. As a name for the inscription of human—industrial or agricultural—activity upon the stratigraphic record, the "Anthropocene" often gives short shrift to other kinds of deliberate, collaborative human-Earth print-making such as the network of Choctaw earth mounds which Chadwick Allen discusses in his current book in progress *Earthworks Rising: Mound Building in Native Art, Literature, and Performance*. A portion of this book has appeared here: "Re-scripting Indigenous America: Earthworks in Native Art, Literature, Community." *Twenty-First Century Perspectives on Indigenous Studies: Native North America in (Trans)Motion*. Ed. Birgit Daewes, Karsten Fitz, and Sabine N. Meyer (New York and London: Routledge, 2015).

4 The wind is Andreas Malm's privileged figure for a type of temporal energy that, like water, cannot retain what it releases, and that resists privatization to the extent that it does not lend itself as easily as carbon to being stocked or stored. We can compare this evasion of appropriation to Thoreau's describing the wind as "sigh[ing] over the fields, which no shrub interprets into its private grief." See also Branka Arsic, *Bird Relics.*

5 In *Timescapes of Modernity*, Barbara Adam takes radiation, not climate change, as her primary example of a phenomenon whose

> materiality falls outside the conventional definition of the "real," outside conceptions where real means material and where this in turn is defined by its accessibility to the senses. Invisibility, vast, incredibly fast and variable time-spans of decay, networked interdependence and the fact that the effects are not tied to the time and place of emission, therefore, make radiation a cultural phenomenon that poses problems for traditional ways of knowing and relating to the material world.
>
> *(10)*

For readers of Rachel Carson's *Silent Spring*, this latency and deferral of outcome, whereby symptoms appear long after their initiating causes, define environmental relations in general. On climate change as a structure of deferral, see also Srinivas Aravamudan's "The Catachronism of Climate Change."

6 As Ana Baginski, a PhD candidate at UC-Irvine, suggests to me, "the fantasy of constant receptivity might be an effect of not being able to tell what or when receptivity is."

7 www.rawstory.com/2016/06/unprecedented-scientists-declare-global-climate-emergency-after-jet-stream-crosses-equator/

8 For a critique of the notion of "Indian time" as a measure of American Indian "authenticity," see Cheryl Wells, " 'Why[,] these Children Are Not Really Indians': Race, Time and Indian Authenticity," *American Indian Quarterly* 39, no. 1 (Winter 2015): 1–24. That attention to non-clock-based temporal cues should get coded as "authentic," "timeless," or "in touch with nature" is one measure of the flattening perspective of clock-time.

9 Anderson devotes the last third of her book to ongoing efforts to revive these methods of harvest and management.

10 I quote here from Annie Montaut's translation of Mishra's Hindi into French; the book has not yet, as far as I know, been translated into English.

11 For Mishra, the richness of vocabulary for water-related phenomena in the language of the region indicates that people remember the desert was once ocean and is in a sense still composed of water (49).

12 For comparable critiques of the way in which agriculture institutes the metaphysical illusion of presence as "constant presence"—an illusion dependent on the perpetual clearing away of objects life-forms deemed foreign to the field—see Timothy Morton (2016, 48) and Colin Duncan (1996, 46).

13 In early 2016 a group of astronomers at the University of Texas, Arlington, published a paper claiming to have "dated the poem"; the paper has since been retracted.

14 David Campbell's translation reads: "The moon has set, and the Pleiades; it is midnight, and time goes by, and I lie alone" (173). It is with some trepidation that I venture to write as a non-classicist about this contested fragment whose lyricism, scholars have claimed, argues against its attribution to Sappho, even if—and precisely because—the figure of the lonely, feminized self speaking to itself at night accords too well with the belated reconstruction of "Sappho" as "lyric" analyzed by Yopie Prins in *Victorian Sappho*

(Princeton, NJ: Princeton University Press, 1999). The following reading should be taken as a speculative elaboration on Castoriadis's Fragment 168b, not Sappho's per se. For a history of the debate over whether the lines belong to Sappho, see, for example, Diskin Clay (1970) and Paula Reiner and David Kovacs (1993).

15 I am drawing here from M.L. West's pellucid explanation of a star's heliacal rising and setting in his "Excursus II Time-Reckoning" appended to his edition of Hesiod's *Works and Days* (New York: Oxford University Press, 1978, 379). His recourse to the future perfect speaks volumes about the play of anticipation and retrospection necessary to the date's determination.

16 Carson does not translate the repeated δέ that cuts as much as it joins the different actors of each line, even as a Keatsian "still" might. Clay argues it is into Endymion's bed that Selanna has set, suggesting again both greater continuity and separateness between the human "ego" and the celestial bodies than may be apparent to modern readers (124).

17 I borrow here from Christopher Childers's recent translation: "as, after the sun has dipped his flaring head,/the moon's rose fingers brush the air//and shutter all the stars, and an equal luster/shimmers the ocean's salty surf/and lights the meadows where the flowers cluster,//where the dew scatters spangles and the rose/riots, where melilot uncloses/blossoms and, delicate, the chervil blows." For a beautifully pertinent discussion of nineteenth-century American beliefs in the restorative powers of lunar light, see Arsic's *Bird Relics*.

18 While some contemporary gardening manuals and lunar calendars recommend harvesting medicinal herbs under a waxing moon because their "oils are more concentrated at this time," others prioritize the cycle of ascent and descent. It's beyond the scope of this essay to assess the scientific merits of what is often roundly dismissed as pseudoscience, but any such assessment would have to start by discussing the long history of ideological dismissals of folk and indigenous ways of knowing in the constitution of science as a field.

19 While biodynamic farming, which follows one or more lunar calendars, is usually associated with the formal teachings of Rudolf Steiner, numerous proverbs across different cultures bear witness to long-standing practices of timing certain actions—planting seeds, cutting trees, and storing apples—in relation to lunar cycles. It is the ambition of the larger project of which this essay is a part to recover some of this archive and unblock Steiner's hold. The different names given to successive moons by different American Indian tribes is a topic of recent popular attention; here too it is important to respect the specificity and changing nature of such naming practices.

20 Here is Rei Terada in a post from November 2016 giving us the terms by which to understand the drought as a protracted, slow-burning fire making room for the flowers that follow:

> if the first part of settler colonial thinking clears away inhabitants because they are there, the second part involves claiming that there's nothing left. I often feel I should be taking more photos of dead trees, and I hope someone is, yet also the forest belongs to the shrubs and flowers now and they know what they're doing. They're pushing drought tolerance to new levels. They're enough to hold the earth on the rock and condition the soil, and everything that can be 'done about' the forest, and whatever it can become in future, is being done by them, if they're left alone to do it. Compared with that, the articles underlining the devastation are settler colonialism phase two rhetoric, preparing the case for 'brush clearing'

> in the name of human fire safety (therefore more erosion and more fire) and the finishing-off that makes way for commercial development.

For a related critique of the logic of emergency determining contemporary firefighting, see also www.yesmagazine.org/planet/when-fighting-wildfires-does-more-harm-than-good-20161206

References

Adam, Barbara. 2010. "History of the Future: Paradoxes and challenges." *Rethinking History: The Journal of Theory and Practice* 14 (3): 361–78. doi:10.1080/13642529.2010.482790.

———. 1998. *Timescapes of Modernity*. New York: Routledge.

Anderson, M Kat. 2005. *Tending the Wild: Native American Knowledge and the Management of California's Natural Resources*. Berkeley: University of California Press.

Arsić, Branca. 2016. *Bird Relics: Grief and Vitalism in Thoreau*. Cambridge, MA: Harvard University Press.

Aravamudan, Srinivas. 2013. "The Catachronism of Climate Change." *Diacritics* 41 (3): 6–30. doi:10.1353/dia.2013.0019.

Bauman, Zygmunt. 2000. *Liquid Modernity*. Malden, MA: Blackwell.

Bonneuil, Christophe and Fressoz, Jean-Baptiste. 2015. *The Shock of the Anthropocene: Earth, History, and Us*, translated by David Fernbach. New York: Verso Books.

Bruyère, Vincent. 2017. "Ice Cold Grief: 1610 and the Orbis Hypothesis." Paper presented at the annual meeting of the American Comparative Literature Association, Utrecht, Netherlands, July 6–9.

Burnett, Anne Pippin. 1983. *Three Archaic Poets: Archilochus, Alcaeus, Sappho*. Cambridge, MA: Harvard University Press.

Campbell, David. 1982. *Greek Lyric I: Sappho and Alcaeus*. Cambridge, MA: Harvard University Press.

Carson, Anne. 2003. *If Not, Winter: Fragments of Sappho*. New York: Vintage Books.

Castoriadis, Cornelius. 2007. *Figures of the Thinkable*, translated by Helen Arnold. Stanford, CA: Stanford University Press.

Childers, Christopher. 2015. "TRANSLATION – Sappho, Horace, Heraclitus, Alcaeus, and Callimachus." Last modified October 22, 2015. Accessed April 17, 2019. http://columbiajournal.org/translation-5-poems-by-christopher-childers.

Clay, Diskin. 1970. "Fragmentum Adespotum 976." *Transactions and Proceedings of the American Philological Association* 101: 119–129. doi:10.2307/2936043.

Crary, Jonathan. 2013. *24/7: Late Capitalism and the End of Sleep*. New York: Verso.

Crutzen, Paul and Eugene Stoermer. 2000. "The Anthropocene." *IGPB (International Geosphere-Biosphere Programme) Newsletter* 41 (May): 17–18.

duBois, Page. 1995. *Sappho is Burning*. Chicago: University of Chicago Press.

Duncan, Colin. 1996. *The Centrality of Agriculture*. Buffalo, NY: McGill-Queen's University Press.

François, Anne-Lise. 2017. "Ungiving Time: Reading Lyric by the Light of the Anthropocene." In *Anthropocene Reading: Literary History in Geologic Times*, edited by Tobias Menelyand Jesse Oak Taylor, 239–258. University Park, PA: Pennsylvania State University Press.

Haraway, Donna. 1991. *Simians, Cyborgs, and Women: The Reinvention of Nature*. New York: Routledge.

Hesiod. 1914. *Hesiod, the Homeric hymns, and Homerica*, translated by Hugh G. Evelyn-White. Cambridge, MA: Harvard University Press. Perseus Digital Library. http://data.perseus.org/citations/urn:cts:greekLit:tlg0020.tlg002.perseus-eng1:1–10

Lewis, Simon L. and Mark A. Maslin. 2014. "Defining the Anthropocene." *Nature* 519: 171–80. doi:10.1038/nature14258.

Luciano, Dana. 2015. "The Inhuman Anthropocene." *Avidly: A Channel of the Los Angeles Review of Books*. Last modified March 22, 2015. Accessed April 17, 2019. https://avidly.lareviewofbooks.org/2015/03/22/the-inhuman-anthropocene/.

Malm, Andreas. 2016. *Fossil Capital: The Rise of Steam Power and the Roots of Global Warming*. New York: Verso.

Mishra, Anupam. 2000. *Traditions de l'eau dans le désert indien: les gouttes de lumière du Rajasthan*, translated by Annie Montant. Paris: L'Harmattan.

Morton, Timothy. 2016. *Dark Ecology: For a Logic of Coexistence*. New York: Columbia University Press.

Reiner, Paula and Kovacs, David. 1993. "ΔΕΔΥΚΕ ΜΕΝ Α ΣΕΛΑΝΝΑ: The Pleiades in Mid-Heaven (*PMG* Frag. Adesp. 976 = Sappho, Fr 168 B Voigt)." *Mnemosyne*. XLVI, Fasc. 2: 145–159.

Shaw, Evelyn and Darling, Joan. 1985. *Female Strategies*. New York: Walker and Company.

SECTION 2

The subject of climate

4

INDIGENOUS REALISM AND CLIMATE CHANGE

Kyle Powys Whyte

Climate realisms

The concept of anthropogenic (human-caused) climate change is not profoundly new for many of the Indigenous persons I know. In this chapter, I'll build on this thought to outline an *Indigenous realist* method for understanding the connection between human societies and environmental change. As a short chapter, my points will be rather brief and theoretical, covering quite a few concepts rapidly. I'm going to just start with the content of this chapter right now, instead of having an elaborate introductory section foreshadowing every forthcoming point I'll be making.

To begin with, scholars from my own Tribe and community (Potawatomi) have certainly noted that anthropogenic climate change is not profoundly new. Robin Kimmerer, for example, has said for many years that "Once again, we are in a situation of forced climate change adaptation" (Kimmerer 2014). Kimmerer can be read as invoking at least four contexts here when I interpret their words. First, in Potawatomi and many other Indigenous traditions, our oldest philosophies consist of stories, conundrums, debates, theories, and methodologies of how societies can be organized to adapt to environmental change of different types, including weather, seasons, and climate. They are about how diverse plants, animals, physical entities, insects, biota, and flows can relate to one another responsibly under conditions of constant change. Through practices such as burning, Anishinaabe people sought to influence landscapes in ways that supported interdependent relationships between humans and diverse beings and entities beyond humans. By forming different organized bodies (smaller and larger) throughout the year, people moved seasonally across the region to ensure they were in the best places for tending ecosystems, harvesting, storing valuables, engaging in ceremony, conducting trade and diplomacy, and fostering kin relations. Critically, our philosophies about these ways of life focus on understanding and living with the challenges of flourishing under conditions of constant change. Far from focusing on the nature of property, rights, or contracts, our traditions are about figuring out—in complex ecological systems—what responsibilities humans and nonhumans have to each other and the best ways to organize society to ensure everyone is motivated to carry out their responsibilities, revise them when needed, and pass enduring responsibilities to future generations. Here I am not making the point that Potawatomi peoples' traditions have the answers to sustainability and resilience; what I'm saying is that our oldest philosophies are about connections between morality (especially shared responsibilities) and environmental change.

Second, Potawatomi and Anishinaabe peoples, and we are not alone in this, have ancient stories of large-scale migration, in which our ancestors had to move drastically across different climate regions and learn how to live (again) responsibly in those new places. Deborah and Hilary McGregor (2018) have recently discussed how many of our ancient stories are about climate change.

Third, Kimmerer is referring to a recent case in the nineteenth century in which the U.S. settler society forced many Potawatomi people to relocate from the Great Lakes region (our homelands) to Kansas and then to Indian territory (now Oklahoma). These were human-caused changes regarding moving from one climate region to another. At the time (and since), the United States imposed deeply "un-adaptive" measures on Potawatomi people, including how settlers generated the conditions for the "dust bowl" drought period in the 1930s, forced Indian people to become farmers, and took diverse measures to reduce the size of Tribal lands and support extractive industries. In this way, U.S. colonialism inflicted harmful and risky anthropogenic environmental and climate change on Indigenous peoples. This point can be understood whether by environmental and climate change we mean ecological changes, such as transforming soil composition or forest cover, or social changes that alter how humans can adapt to the dynamics of ecosystems, such as forcing people to farm a certain way or dispossessing them of their lands.

Fourth, Tribes like the Potawatomi in Oklahoma and the Great Lakes region are now facing climate change again, as changes in precipitation patterns, warming and extractive industries are engendering further harms and risks. These harms and risks threaten our capacities for the continuation of our cultures without unnecessary disruptions, the thriving of our economies, and our political rights as sovereign peoples.

Daniel Wildcat, in his 2009 book, *Red Alert! Saving the World Through Indigenous Knowledge*, refers to a method of interpreting environmental change that I relate to the types of Indigenous traditions I just referenced for Potawatomi people:

> In North America many indigenous traditions tell us that reality is more than just facts and figures collected so that humankind might widely use resources. Rather, to know "it"—reality—requires respect for the relationships and relatives that constitute the complex web of life. I call this indigenous realism, and it entails that we, members of humankind, accept our inalienable

> responsibilities as members of the planet's complex life system, as well as our inalienable rights.
>
> *(xi)*

Indigenous realism, according to Wildcat, involves an interpretive framework for understanding environmental change that highlights relationships, relatives, responsibilities and interdependence ("the web of life"). Instead of focusing exclusively on resource distribution, facts, figures, and rights, Wildcat argues for emphasis on the moral connections that are part of the "complex web of life." Interpreting Kimmerer's work, I did my best to show how the context they are implying are relational contexts. The relational contexts are stories of connections involving responsibility and land—whether in terms of the responsibilities of seasonal living or how colonialism disrupts people from exercising their responsibilities. In some more ancient traditions, there are times when the relationships involved myriad responsibilities connecting humans, plants, insects and diverse others interdependently both locally and regionally. In some ancient time periods, we look to the flourishing of systems of responsibilities as an ideal toward which we can aspire today—in our own ways. In more recent traditions, unfortunately, many of the relationships are lacking responsibilities across different human and nonhuman groups. Systems of power—especially colonialism, capitalism, industrialization, patriarchy, and extraction—are major factors causing environmental harms and risks for current and future generations. I see little evidence that any of these systems are based on mutual responsibility as a core operating philosophy. They certainly do not prioritize mutual responsibility. Hence, one of the emphases of Indigenous realism is to focus analysis of the present in relation to the current state of interconnected responsibilities. This emphasis can generate analyses of existing situations that reveal how recent problems of mutual responsibility are tied to past precedents where mutual responsibilities were unraveled.

Indigenous realism can be contrasted with one grounded instead in the language of a certain kind of empiricism. As Wildcat points out, "facts and figures" and "resource distribution" are very much emphasized in the portrayal of climate change in recent academic and media reports. The Intergovernmental Panel on Climate Change issued a special report in 2018 showing the economic advantages of keeping warming at 1.5°C as opposed to 2°C. In the summary for policy makers that is available online, the report states that "Climate-related risks to health, livelihoods, food security, water supply, human security, and economic growth are projected to increase with global warming of 1.5°C and increase further with 2°C." The goal of the report is to show that "Pathways limiting global warming to 1.5°C with no or limited overshoot would require rapid and far-reaching transitions in energy, land, urban and infrastructure (including transport and buildings), and industrial systems (high confidence)" (IPCC 2018, A.3.). The IPCC report characterizes climate change in sobering scientific language, and does so to suggest a stern warning to humans. The warning is that humans need to change their behavior, as the facts and figures tell us that resources will be further maldistributed today and in the future for some populations. Resources refer to goods such as access to water, health, economic opportunity, among others listed in the quote from IPCC.

In media coverage of climate change in the same year, major outlets covered the U.S. Fourth National Climate Assessment. They made equally stern warnings about how climate change will precipitate futures involving billions of dollars in economic losses and thousands of deaths. CNN, for example, phrased the report as "a dire warning about climate change and its devastating impacts, saying the economy could lose hundreds of billions of dollars—or, in the worst-case scenario, more than 10% of its GDP—by the end of the century." It goes on to state that "The costs of climate change could reach hundreds of billions of dollars annually" and the "The Southeast [U.S.] alone will probably lose over a half a billion labor hours by 2100 due to extreme heat" (Christensen and Nedelman 2018). This story discusses in detail findings from the report pertaining to major harms and risks of climate change to peoples' health and the economy. Again, the coverage conveys the idea that were the United States to "immediately reduce its fossil fuel use and greenhouse gas emissions, it could save thousands of lives and generate billions of dollars in benefits for the country." Similarly, widely distributed news outlets, such as the *New York Times*, continue to run stories about how particular climate change impacts, such as intensified droughts, are ruining communities and lives. A scan of 2018 titles include "A Warming World Creates Desperate People" in the *New York Times* (Markham), and "Climate change swells ranks of refugees as Trump administration retreats to the sidelines" (Wilkinson) in the *Los Angeles Times*. This way of accounting for harms and risks is often portrayed as a type of realism about climate change. The realism is credited as being based on the very best scientific evidence and the utmost concern for the well-being of humans and nonhumans into the future.

The purpose of this genre of climate realism is both to motivate people everywhere to take action to address climate change and to counteract the different types of skepticism about the importance of investing in measures to mitigate and adapt to climate change. Whether such realism has been effective has long been in question by a number of scholars. Kari Norgaard's study, *Living in Denial* (2011), challenges the idea that people are motivated to act on climate change by having access to more vivid information or who have a competent scientific understanding. Norgaard includes extensive references of scholarship challenging what I am calling climate realism here. Elizabeth Arnold (2018) has critiqued the ways in which a focus on environmental impacts makes groups' suffering from immediate climate change impacts "helpless and voiceless." Emilie Cameron (2012) argues that certain accounts of climate vulnerability for small communities mask the continuation of extractive industries in relation to those same communities. Elizabeth Marino (2012) shows how focuses on coastal erosion obscure that the factors producing vulnerability are quite old, traceable to colonialism. These different scholars, in diverse ways, demonstrate how climate realism ignores systems of power and historical precedent in descriptions of harms and risks that focus on climate change impacts themselves as the major causal factors.

Indigenous perspectives on climate change

Indigenous leaders and scholars have, for generations, focused on how the language used to identify environmental phenomena can be articulated to avoid taking responsibility for oppression. In formal international and academic settings, since *at least* the 1990s, Indigenous leaders, scholars, and journalists have been among those offering perspectives and philosophies that challenge this way of thinking about climate change. Hence, by the time Indigenous peoples enter formal climate change discussions at national and international scales, they already had been developing contemporary versions of accounts of the relationship between climate change and responsibility, flowing from the ancient philosophies and contemporary experiences I had described earlier. In 1994, the College of Menominee Nation founded its Sustainable Development Institute as a research and educational center for drawing lessons from Menominee culture and experiences resurging against colonialism (Dockry et al. 2016). In 1997, Andrea Tunks published an article covering Māori philosophies of climate change rooted in ancient stories and relationships of responsibility and care, writing that "climate change is not a new phenomenon" yet "this time, the damage may be worse" (83). A 2002 meeting between diverse Pacific and North America Indigenous peoples led to the creation of the United League of Indigenous Nations, which includes a major provision on "Protecting our Indigenous lands, air and waters from environmental destruction resulting from global warming through exercising our rights of political representation" (Grossman 2008).

In 1998, 180 Indigenous delegates met in the traditional territory of Pueblo peoples (New Mexico) for the "Circles of Wisdom" Native People/Native Homelands Climate Change Workshop. The Albuquerque Declaration, published based on the workshop, referenced directly Indigenous concerns and philosophies about climate change. It says that "climate imbalance will cause the greatest suffering to the Indigenous peoples" and that fossil fuel development has "Cultural impacts, forced removal, land appropriation, destruction of sacred and historically significant areas, breakdown of Indigenous social systems, and violence against women and children" (Maynard, 1998, 71–72).

Critically, the chapter related to the workshop connects harms and risks to Indigenous peoples with previous forms of environmental change:

> Over the last 500 years, essential environmental balances that had sustained Native peoples in North America for many millennia began to rapidly shift. Forests were cut for homesteads and farming. Alien plants displaced grasslands. Dry lands flooded, rivers changed their courses, and ponds and swamps drained away as watercourses were dammed and channeled. Important

> providers of nourishment and protection—buffalo, salmon, eagle, wolf, and shad—were pushed to near extinction. New and strange creatures—horse, cow, pig, sheep, and pheasant—shoved aside indigenous species and came to dominate local economies. Exotic new diseases eradicated whole villages. Tribal social, political, cultural, and spiritual relationships throughout entire regions collapsed. Spiritual leaders lost their followers. Communities—even entire Tribal nations—were extinguished or forced to relocate.
>
> *(Houser et al. 2001, 355)*

The chapter goes on to discuss how colonialism was an infliction of climate change on Indigenous peoples.

> Native peoples experienced continental scale changes in their surroundings that are not unlike the types of changes that all Americans, indeed, all peoples may face in coming decades. The changes were substantial in magnitude, surprising in their occurrence, unmanageable by available technologies and existing forms of government, and irreversible. In those respects, the changes may provide insights of the kinds of transformations—cultural, economic, and social—that global changes in climate may bring, both for Native peoples and for America as a whole.
>
> *(355)*

This work was part of the first U.S. National Climate Assessment in 2001. The report and chapter point to the following ideas that I would associate with Wildcat's concept of Indigenous realism: (1) that Indigenous peoples have longstanding intellectual traditions about how to negotiate environmental change, (2) that climate change is bound up with colonialism, capitalism and industrialization, and that (3) the concept of climate change can mask the need to address systems of power.

In Candis Callison's book, *How Climate Change Comes to Matter* (2014), she discusses interviews with Indigenous scientists and leaders who had been working since the 1980s and 1990s on climate change, including Patricia Cochran and Sheila Watt-Cloutier. Cochran stated that:

> Certainly, when our elders talk about climate change and global warming, those are not the words that anybody would ever hear coming form an elder's mouth or anybody else. Maybe because those are just not the words that we use. But if you were to ask elders about the changes in ice conditions, and what they have seen in their lifetime, changes in ice? Well, that would be a three-month conversation [...] All things are connected, and so to take one piece of a problem and not connect it to the rest of the world and the environment around? It just logically makes no sense. How can we talk about changes in weather without talking about changes in vegetation or the air or the people or the animals, as all those things are part of a natural mix.
>
> *(Callison 2014, 45–46)*

Cochran critically points out that based on Indigenous histories and experiences, people are observing environmental change, and associating it with the factors that are salient to their ways of knowing the world. For whom anthropogenic climate change is not a profoundly new topic, their sense of "realism" is not going to be tied to the discursive frame I described in the previous section.

Callison references Sheila Watt-Cloutier as saying that "I think that some people have not fully come to understand that there is no disconnect between the suicide rates in our communities and climate change (Callison 2014, 67). Callison describes Watt-Cloutier and her colleagues' efforts at the Inuit Circumpolar Council (ICC), crystalized in 2004, to advance a petition on climate change and human rights to the Inter-American Court. The court rejected the petition in 2006, and some people have observed that the strong presence of the United States as an influencer of the court may have played a big role in that rejection. Callison writes that

> Despite the failure of the case and the ways in which it largely fell out of mainstream media coverage, the petition moved the experience of climate change outside of the realm of mere illustration and into the domain of self-determination, power relations, and settled causality. In other words, it isn't greenhouse gas emissions doing this to the Inuit, but the U.S. government, which has stalled on mitigation policies that might prevent further loading of emissions such that climate change will occur more precipitously at the poles, where effects are known to be more extreme.
>
> *(67)*

For Callison, Watt-Cloutier's "focus on connection doesn't do away with questions about scientific uncertainty, but makes them somewhat irrelevant. It instead evokes the ideal of precaution and communality as well as moral and ethical responsibility" (73). For Callison, in her dialogues with Watt-Cloutier and Cochran, the concept of "climate change" can be seen as a distraction when it is isolated from Indigenous histories and experiences and with systems of power that are major causal factors behind harms and risks. Regarding the Native Peoples/Native Homelands movement or Callison's ethnographic study, among other references in this chapter, it is not surprising why Wildcat phrases Indigenous realism as a method that focuses on relationships, responsibilities, and the complex web of life. For it seems that Indigenous scholars and leaders have directly interpreted climate change most immediately through its connection to relationships between humans and nonhumans, which is a different approach than that of asserting facts and figures or risks of misdistributions of resources.

Indigenous realism and colonialism

One of the key aspects of Indigenous realism that I want to bring out is the direct engagement with systems of power in relation to the environment and climate change. Here I will focus on colonialism. From an Indigenous realist perspective, colonialism can be understood as a system of power that is based on the infliction of anthropogenic climate change on different groups of people. If, in extremely broad strokes, capitalism is a system of economic exploitation, then we might see colonialism as referring to a system that undermines colonized peoples' capacity to adapt to the invasion, establishment of extractive industries and/or settlement of other societies. As capitalism and colonialism worked together in contexts like the United States or Canada, I think a lot of Indigenous peoples understand colonial processes as laying the groundwork for and protecting the maintenance of capitalist industries. These industries include coal mining and oil drilling, large-scale agriculture, deforestation, and the creation of large urban areas—in short, the drivers of today's ordeal with anthropogenic climate change. Disingenuous and bad faith approaches to treaty-making sought to remove Indigenous peoples from the majority of their lands. Boarding schools and other problematic forms of education strip Indigenous peoples of languages that express knowledge and skills related to particular ecosystems, seasonal change and knowledge. Forms of political recognition, such as U.S. Indian Reorganization Act of 1934 or the Alaska Native Claims Settlement Act of 1971, created Tribal governments and Alaska Native corporations that seek to redefine Indigenous peoples' governments as capitalist enterprises whose goal is to mimic the U.S. economy by investing in carbon-intensive economic activities, including Tribes that are heavily involved in coal mining. What is insidious about the fact that Indigenous peoples are concerned about "climate" related harms and risks today is that the same colonial and capitalist institutions that fostered carbon-intensive economic activities on Indigenous territories are still to blame. In this way, climate injustice against Indigenous peoples refers to the vulnerability caused by ongoing, cyclical capitalism and colonialism. For capitalist and colonial institutions facilitate carbon-intensive economic activities that produce adverse impacts while at the same time interfering with Indigenous people's capacity to adapt to the adverse impacts.

Indigenous realism situates today's climate change "crisis" *systematically* among the different historic and contemporary factors that lead to what becomes labeled "climate" related harm and risk. Wildcat, for example, argues that it is important to situate climate-related resettlement of Indigenous peoples living in coastal areas with the longer history of *displacements* that these peoples were also not responsible for causing. The displacements include "geographic" (displacement, e.g. Trail of Tears and removal to reservations), "social" and "psycho-cultural" (such as through removal of children to boarding schools) (Wildcat 4). Callison writes that attention must be paid to what "climate change portends for those who have endured a century of immense cultural, political and environmental changes" (Callison 2014, 42). Sheila Watt-Cloutier, discusses in an interview how "Climate change is yet another rapid assault on our way of life. It cannot be separated from the first waves of changes and assaults at the very core of the human spirit that have come our way" (Robb 2015; Watt-Cloutier 2015).

In this way, Indigenous realism, as an approach, is skeptical about any claims that some group of Indigenous peoples is "the first" climate "refugees"—as several media stories have claimed. The term "first" is problematic because it erases the layered histories of how displacement has been inflicted on Indigenous peoples by colonialism. Forced displacements are almost always forms of anthropogenic climate change, since they involve one group's infliction of environmental and climatic change on another group, as I covered earlier in this chapter. In terms of refugees, this term is problematic for masking two realities. First, Indigenous peoples who are adapting to climate change are often times doing so in regions that are part of their ancestral homelands. Their ancestors did not want to give up their access to those lands and waters. Second, it masks how for many Indigenous peoples they are already living in diaspora, and seeking to rebuild their political, cultural, and economic lives.

This approach is used in the research and educational programs discussed by Megan Bang et al. (2014). In one program, they take on the challenges of the term "invasive" as it is used in conservation education in the city of Chicago.

> A particular content focus we took up was the importing of plants from other places and specifically common Buckthorn, a native species in Europe that was brought to North America in the early 1800s. This plant is particularly destructive to woodlands and oak savannahs and is considered a deeply problematic invasive species. The act of naming became particularly important as we continued to develop curricula around "invasive species" (see Bang and Medin 2010; LaDuke 2005). We, the teachers in the program, recognized our use of the term invasive species signaled a particular epistemic and ontological stance to youth—a western science one specifically—and not one that we intended. Thus, the term invasive species placed buckthorn, and other plants that were forcibly migrated to Chicago, outside our design principle around naming our plant relatives because while they may not have been our relatives, the term disposed them as relatives to any humans. Further, the term failed to make visible the motivation of settlers that brought flora and fauna from their homelands to make these new lands like home—or what has been termed ecological imperialism (Crosby 2004; McKinley 2007, 11).
>
> *(Bang, Curley, Kessel, Marin, Suzukovich III and Strack 2014, 47)*

By "ontological" and "epistemic," Bang et al. refer to how the term "invasive," depending on how it is used, affects how we experience (ontology) and come to know (epistemic) the origins of the different beings we are related to in a shared environment. "Invasive" can mask systematic analysis of the factors leading up to the recent presence of a plant in a place and the factors that make it possible for the plant to be deemed threatening.

As Audra Mitchell's (2016) research shows, today's global discourses of extinction are often so focused on "species" that they cannot come to grips with Indigenous peoples' experiences of having their relationships with nonhumans greatly disrupted by colonialism. Mitchell's point corresponds to my many conversations in the last several years with Preston Hardison, a policy analyst for the Tulalip Tribes and advocate of the protection of Indigenous knowledge, we often discuss how for many Indigenous peoples, the loss of local access to a culturally or economically significant plant or animal due to colonial domination is comparable to that species becoming extinct (Hardison, *personal communication*). "Extinction," then, is also a term that suggests a particular way of experiencing and coming to know the world. Within this method, then, a structure of power can contribute to climate change through the physical machinations of how it imposes its own aspirations on other people. As is the case in settler colonial studies, the structure involves using terms that mask ways of experiencing and ways of knowing, as Bang et al. discussed. Hence, Leanne Simpson writes how "Indigenous peoples have always been able to adapt, and we've had a resilience. But the speed of this—our stories and our culture and our oral tradition doesn't keep up, can't keep up … Colonial thought brought us climate change" (Simpson quoted in Klein 2013). I read this passage as suggesting both how colonial power produces climate change at the same time it invents "climate change" as a topic that obscures factors leading to harm and risk.

I remember recently seeing coverage of how the first image of the planet earth from space was one of the motivators of environmentalism in the 1960s and 1970s in countries like the United States. Similarly, in the IPCC and NCA, there is a concern about protecting *this* world from becoming hotter than 1.5°C in global average temperature. In cases of climate resettlement, the stories often portray the place where people are considering moving away to as *the* homeland. *This world* that is worth protecting from environmental catastrophe is a world of terms such as *anthropogenic climate change*, *invasive species*, *extinction*, and *the first climate refugees*. It is not surprising that these terms are used, of course. Terms like these must be used so that it is possible to think there is anything about *this world* that is worth protecting. For the current state of affairs of *this world* is one in which those ecosystems that are sought to be saved are ones that emerged through the destruction of Indigenous homelands. Mitigating invasive species and stopping extinctions does little to address that the places *so protected* are ones that our Indigenous ancestors would have seen as dystopian spaces.

Indigenous leaders and scholars have referred to today's times as ones are ancestors would have seen as "dystopian" (Lee Sprague, *personal communication*), "apocalypse" (Larry Gross, [Gross 2016, 33]), and "post-Native Apocalypse" (Grace Dillon [(Dillon 2012, 10]). Building on Dillon's research, Conrad Scott's (2016) recent study discusses how "Indigenous literature, following the culturally destructive process of colonial European advancement and absorption of what are now called the Americas, tends to narrate a sense of ongoing crisis rather than an upcoming one" (77). Indigenous peoples then do not always approach the climate crisis as an impending future to be dreaded. Heather Davis and Zoe Todd see an insidious irony in the different ways Indigenous and nonIndigenous persons approach the anthropocene and climate crisis. They describe colonialism as a seismic shockwave that "kept rolling like a slinky [as it worked] to compact and speed up time, laying waste to legal orders, languages, place-story in quick succession. The fleshy, violent loss of 50 million Indigenous peoples in the Americas is something we read as a 'quickening' of space-time in a seismic sense" (Davis and Todd 2017, 771–772). They then point out that "the Anthropocene or at least all of the anxiety produced around these realities for those in Euro-Western contexts—is really the arrival of the reverberations of that seismic shockwave into the nations who introduced colonial, capitalist processes across the globe in the first half-millennium in the first place" (Davis and Todd 2017, 774).

Returning to some previous themes, Indigenous realism is a method for analyzing climate change that focuses on relationships and responsibilities. The authors referenced in this section point out how discourses oriented around facts and figures actually fail to tell the stories of relationships responsibilities, and histories embedded within complex and interdependent webs of life. Terminology like "invasive" species or phrases that suggest it is "climate change" that is forcing people to move, ignores the systems of power that have still not been named and addressed, and that play a larger role in shaping why Indigenous peoples and other groups are concerned in particular about *today's* climate change ordeal. It is not that the impacts themselves are surprising or all that different from generations of colonial violence. Rather, it is more about the horror of watching current generations of persons, whether in academia, media, or politics, seek once again to erase their own responsibility to name and address the systems of power that continue to affect Indigenous peoples. Indigenous realism is about establishing and repairing the relationships of responsibility that can support flourishing human and nonhuman lives who face constant environmental change. Importantly, many of the people I have cited work actively in climate science and public media. So Indigenous realism does not exclude the "facts and figures" based approaches or the importance of bringing urgent attention to harms and risks. Rather, Indigenous realism calls for attention to what is absent in so much climate change discourse, and seeks to draw on Indigenous traditions and philosophy and lessons learned from enduring colonialism to support better futures for the coming generations.

References

Arnold, Elizabeth. 2018. "Doom and Gloom: The Role of the Media in Public Disengagement on Climate Change." Shorenstein Center on Media, Politics and Public Policy. Last Modified May 29, 2018. https://shorensteincenter.org/media-disengagement-climate-change/.

Bang, Megan, Lawrence Curley, Adam Kessel, Ananda Marin, Eli S. Suzukovich III, and George Strack. 2014. "Muskrat Theories, Tobacco in the Streets, and Living Chicago as Indigenous Land." *Environmental Education Research* 20 (1): 37–55.

Bang, Megan and Doug L. Medin. 2010. "Cultural Processes in Science Education: Supporting the Navigation of Multiple Epistemologies." *Science Education* 94 (6): 1008–1026.

Callison, Candis. 2014. *How Climate Change Comes to Matter: The Communal Life of Facts*. Durham, NC: Duke University Press.

Cameron, Emilie S. 2012. "Securing Indigenous Politics: A Critique of the Vulnerability and Adaptation Approach to the Human Dimensions of Climate Change in the Canadian Arctic." *Global Environmental Change* 22 (1): 103–114.

Christensen, Jen, and Nedelman, Michael. 2018. "Climate change will shrink US economy and kill thousands, government report warns." CNN. November 26, 2018. www.cnn.com/2018/11/23/health/climate-change-report-bn/index.html.

Davis, Heather, and Zoe Todd. 2017. "On the Importance of a Date, or, Decolonizing the Anthropocene." *ACME: An International Journal for Critical Geographies* 16 (4): 761–780.

Dillon, Grace L. 2012. *Walking the Clouds: An Anthology of Indigenous Science Fiction*. Tucson, AZ: University of Arizona Press.

Dockry, Michael J, Katherine Hall, William Van Lopik, Cristopher M. Caldwell. 2016. "Sustainable Development Education, Practice, and Research: An Indigenous Model of Sustainable Development at the College of Menominee Nation, Keshena, WI, USA." *Sustainability Science* 11 (1): 1–12.

Gross, Lawrence W. 2016. *Anishinaabe Ways of Knowing and Being*. New York: Routledge.

Grossman, Zoltan. 2008. "Indigenous Nations' Responses to Climate Change." *American Indian Culture & Research Journal* 32 (3): 5–27.

IPCC. 2018. "Summary for Policymakers". In *Global Warming of 1.5°C. An IPCC Special Report on the impacts of global warming of 1.5°C above pre-industrial levels and related global greenhouse gas emission pathways, in the context of strengthening the global response to the threat of climate change, sustainable development, and efforts to eradicate poverty*, edited by V. Masson-Delmotte, P. Zhai, H.O. Pörtner, D. Roberts, J. Skea, P.R. Shukla, A. Pirani, W. Moufouma-Okia, C. Péan, R. Pidcock, S. Connors, J.B.R. Matthews, Y. Chen, X. Zhou, M.I. Gomis, E. Lonnoy, Maycock, M. Tignor, and T. Waterfield. IPCC. www.ipcc.ch/sr15/chapter/summary-for-policy-makers/.

Houser, Schuyler, Verna Teller, Michael MacCracken, Robert Gough, and Patrick Spears. 2001. "Potential Consequences of Climate Variability and Change for Native Peoples and Homelands." In *Climate Change Impacts on the United States*, edited by National Assessment Synthesis Team, 351–378. Washington, DC: US Global Change Research Program.

Klein, Naomi. 2013. "Dancing the World into Being: A Conversation with Idle No More's Leanne Simpson." *Yes! Magazine*, March 5, 2013. www.yesmagazine.org/peace-justice/dancing-the-world-into-being-a-conversation-with-idle-no-more-leanne-simpson.

Kimmerer, Robin W. 2014. "Climate Change and Indigenous Knowledge." Lecture presented at Centre for Aboriginal Initiatives, Centre for Indigenous Studies University of Toronto, March 10, 2014. https://youtu.be/6f7NUi2qhWQ.

LaDuke, Winona. 2005. *Recovering the Sacred*. Boston, MA: South End Press.

Maracle, Lee. 2015. *Memory Serves*. Edmonton, AB: NeWest Press.

Markham, Lauren. 2018. "A Warming World Creates Desperate People." *New York Times*, June 29, 2018. www.nytimes.com/2018/06/29/opinion/sunday/immigration-climate-change-trump.html.

Marino, Elizabeth. 2012. "The Long History of Environmental Migration: Assessing Vulnerability Construction and Obstacles to Successful Relocation in Shishmaref, Alaska." *Global Environmental Change* 22 (2): 374–381.

Maynard, Nancy G., ed., 1998. *Native Peoples-Native Homelands Climate Change Workshop: Lessons Learned*. AGU Fall Meeting Abstracts. Native Peoples-Native Homelands Climate Change Workshop: Final Report. https://digital.library.unt.edu/ark:/67531/metadc11974/m2/1/high_res_d/native.pdf.

McGregor, Deborah, and Hillary McGregor. 2018. "Climate Justice: Listening to the Youth, Listening to the Land." Lecture presented at International Symposium on Indigenous Communities and Climate Change, Princeton University, December 7, 2018.

Mitchell, Audra. 2016. "Beyond Biodiversity and Species: Problematizing Extinction." *Theory, Culture & Society* 33 (5): 23–42.

Norgaard, Kari Marie. 2011. *Living in Denial: Climate Change, Emotions, and Everyday Life.* Cambridge, MA: MIT Press.

Robb, Peter. 2015. "Q and A: Sheila Watt-Cloutier Seeks Some Cold Comfort." *Ottawa Citizen*, May 27, 2015. https://ottawacitizen.com/entertainment/books/q-and-a-sheila-watt-cloutier-seeks-some-cold-comfort.

Scott, Conrad. 2016. "(Indigenous) Place and Time as Formal Strategy: Healing Immanent Crisis in the Dystopias of Eden Robinson and Richard Van Camp." *Extrapolation* 57 (1–2): 73–93.

Tunks, Andrea. 1997. "Tangata Whenua Ethics and Climate Change." *New Zealand Journal of Environmental Law* 1: 67–123.

Watt-Cloutier, Sheila. 2015. *The Right to Be Cold: One Woman's Story of Protecting Her Culture, the Arctic and the Whole Planet.* Toronto: Penguin.

Wildcat, Daniel R. 2009. *Red Alert! Saving the Planet with Indigenous Knowledge.* Golden, CO: Fulcrum.

Wilkinson, Tracy. 2018. "Climate change swells ranks of refugees as Trump administration retreats to the sidelines." *Los Angeles Times*, September 23, 2018. www.latimes.com/nation/la-na-pol-climate-change-refugees-united-nations-20180923-story.html.

5

REALISM'S PHANTOM SUBJECTS

M. Ty

Fragility of realism

To look without eyes that hold one to a vision: this is the peculiar fantasy out of which modern realism has been spun, both in its philosophical and aesthetic variations.[1] One conventional reading of the realist optic—for which the unattended camera might serve provisionally and imperfectly as a figure—interprets such a gaze as the instrument or complement of positivist ambitions. Subjectless, the image of objectivity gleams, unhindered by the phenomenological subject's (passionate) attachment to a specific and inalienable position within the material world. In their most brutal form, hardline realisms, and I quote Trinh Minh Ha's revoicing of Piaget in her critique of the anthropological delusion of detachment, "consist … in ignoring the existence of oneself and [therefore in] taking one's own view for immediately objective and absolute" (Minh-Ha 1989, 55).[2]

As the above remarks may intimate, more often than not realism is regarded as a strong discourse, even among its detractors.[3] This perceived strength is double-fibered: first, realism is attributed with the capacity for generating images of reality whose claim to truth is, to varying degrees, insulated from the contingencies of perception and experience; second, realism is credited with a mimetic potency that not only can render a picture of the world but, like a magic word, can cast it in its image. Arguments for the latter range from theories of fictional and cinematic identification (of which the transformations in German fashion and the emulation suicides following the publication of *The Sorrows of Young Werther* remain a forceful exemplar) to Foucault's anti-humanist account of biopower, which proposes that statistical analyses of the population do not only have a descriptive but also a regulatory function, in that life begins to organize itself "freely" with reference to a mathematically defined range of "normal" variation.[4] As this small sample may show, the assumption that realism enacts or feigns a bivalent mimetic efficacy subtends a wide spectrum of political positions. This power may be met with ideological suspicion or admired for its capacity to confront things "as they really are"—say, in the restrictive world of the bourgeoisie, in the occluded regions of the global periphery, in internal zones of socioeconomic marginalization, in the depiction of the social totality, or in the vista onto a geo-planetary plane of actuality that exceeds the frame of human experience. In other words, considerations of realism—ideological investments notwithstanding—tend to concentrate on its descriptive or constitutive force, and even more often, on the looped point in which they converge. Simplified: the picture of reality is so forcibly real that it is seen to determine reality. In this sense, "realism" signifies less a philosophical movement or aesthetic genre than it embodies the psychoanalytic notion of the reality-principle itself: for the very assessment of reality translates (perhaps all too automatically) into a command to accept and act within the terms of that assessment.

At times, the coherence of realism, understood in the above sense as a strong discourse, is partially authorized by an unspoken, extra-cognitive dictum of tough love—namely, that what is most painful is necessarily most true. "History is what hurts": with a compactness whose ruthlessness drives the message through to compulsory recognition, Jameson's formula is an exemplary instance of the way in which, as Eve Sedgwick observes of the hermeneutic of suspicion, truth-claims sometimes draw their legitimacy from the negative affects with which they are performed (Sedgwick 2003, 102). Realism may, it is true, be inscribed more coolly into an asymptotic approach toward that which speaks for itself—approximating, that is, a dispassionate state of ideally accomplished reference, a zero-degree of mediation that, under a positivist spell, is liable to be conflated with the unflinching felicity of denotative accuracy. But very often, such renderings of representational objectivity ultimately culminate in an encounter with a painful limitation (of subject, identity, world, perception, or mediation), whose denial is presumed to be an efflux of naiveté or idealism. Pain works to testify to truth, whose claim gains solidity by virtue of that very pain. Granted the subtle variations among realisms—each with their specific histories and disciplinary contours—the affective deflation that is characteristic of the "practical" thrust of *realpolitik* is pervasive and, even when associated with the straight assertion of nothing in excess of the "hard facts," is not without ambition to authority, despite claims to be uncolored by an ideological hue.

The appraisal of realism as a truth-discourse propped up by an affirmation of the truth of pain is not intended to underestimate the satisfactions offered by the realist enterprise. Much like the analytic of the sublime, which narrates an overcoming of an experience of perceptual distress by recourse to a compensatory cognitive pleasure, the sting of the real is cut with the gratification of its knowledge. There is a kinship, though not a strict identity, between the "faith in exposure" that sustains paranoid reading and the drive to transparency that animates modern theories of realism (Sedgwick 2003, 130). Of the latter, the promise that Jameson's locates in the ability of the aesthetic to do the work of "cognitive mapping" is illustrative. Taking inspiration from Kevin Lynch, who suggests that alienation within a city arises when subjects are unable to map their own positions in relation to the larger urban whole, Jameson proposes that the aesthetic can similarly assist in the coordination between disparate existential positions and the social totality. In helping to negotiate phenomenological experience with the otherwise invisible global totality to which one is subjected, representation not only has the potential to dispel confusion about the often imperceptible forces that determine lived experience under late capitalism but can also help to allay alienation by way of a "practical reconquest of a sense of space" (Jameson, 1982, 59). Literature, in short, can be orienting. Though Jameson distinguishes such cartographic endeavors from "naive" mimesis—the pellucid mirror that shines an image of the world as it actually is—he does, as he admits, resuscitate for postmodernism something of the spirit of Lukács's faith in realism, which the latter contrasted harshly with potentially frustrating opacities that are characteristic of modernist style (Lukács 1963, 52). One may accept Jameson's claim that cognitive mapping does not aspire to a fully transparent vision of the global "Real" in any simplistic way. Even so, some notion of legibility remains paramount and is the site of investment for the value of the aesthetic, which he understands as fundamentally pedagogical, and hence, political.

If, as one reading of this volume's title may suggest, "climate" were to be taken up an object of realism, such a pairing might seems felicitous in the terms that Jameson outlines—so long as some notion of the "geological" could be successfully substituted for the Marxist horizon of the social (human) totality. Vast, planetary in scope, unfolding over periods of time that may range far beyond the retentive capacities of personal or generational memory, and largely obscured from the perspective of subjective experience even while, in a certain sense, structuring it—anthropogenic climate change poses problematics of apprehension that are shared by the dilemma, familiar to aesthetics, of the subject within a social system too expansive to grasp. One could claim that aesthetic realism is well poised, not only to shuttle between disparate geographic locales but also to manage the scalar shifts that are necessary for grasping the environmental transformations of our moment—shifts, one might add, that are temporal and may leave utterly dwarfed the anthropic frame to which such an act of representation is inevitably bound.

Such an undertaking still, however, would look to realism for its strengths—chief among these, the power to assist the limited subject in overcoming its own provincialism and the attendant promise of cognitive purchase on the potentially overwhelming macrocosm (i.e., the supra-individual, the infrastructural, the ecosystemic, the geo-planetary). Theoretical paradigms that feature realism as a strong discourse take for granted that it is properly located and evaluated with reference to the parameters of epistemology, a domain cut through by a pleasure–pain axis that pertains to different states of knowing. Like Benjamin suggested of "information" as a mode of communication whose truth ultimately comes down to "verifiability," realist representations, under such models of interpretation, are gauged by their (non)correspondence to some plane of extra-representational "objectivity," which, to varying degrees, may be understood to be mediated by language (Benjamin 2005, 147). The will to truth of such discourses can certainly come under scrutiny and be exposed as Eurocentric, positivist, insufficiently dialectical. But such disarmaments of realism's might persist in taking the latter as a premise and target.

What if, somewhat differently, realism could be dislodged from the sublime/oedipal trajectory in which dissatisfaction with debility is superseded by the pained satisfaction of knowledge or else expanded indefinitely into a scene of felt inadequacy? What work might the realist mode perform, apart from the (failed) conquest of that which radically out-scales subjective aspects of the reality of the material world? What of the "climate" may come into view when realism is not looked upon exclusively as a site of the production of a truth-discourse, which is translatable into a blueprint of practical limitations?

This chapter begins by loosening realism from the grip of epistemic coordinates in relation to which it may either be repealed as illusory or affirmed for its verisimilitude, accuracy, "objectivity," or fidelity to life. The following pages consider, alternatively, the *fragility of realism* as a mode of representation whose hold on the world is characterized by a tenuousness that is irreducible to the diagnosis of the (racialized) incapacity of the (uneducable) subject, who tries unsuccessfully to make sense of the world she inhabits.

The heroics and hysterics of the overwhelmed sensorium are but one way to regard the potentially uncontainable referent of "climate." Taken up as mode of representation, realism is denuded of complexity when treated primarily as a static genre. I will suggest, otherwise, that it is more fully comprehended a cultural practice of voyeurism that is drawn toward the asymptotic limit of pure impersonality. The latter, ultimately unattainable aesthetico-philosophical ideal has conventionally been theorized in terms of the secularization of an omnipotent theological presence or as a normative aspiration and fiction that structures the discursive coherence of Western science.

Realism's image-repertoires are read here in a different context, with two threads coming to light. The first traces how problematics of modern realism continues to exert a largely unmarked influence on contemporary conversations about environmental catastrophe. Much of the discourse surrounding climate change, including rhetorics of the Anthropocene, exhibits a tendency toward universalism and an attendant emphasis on "species"—both as a locus of agency and as a transcendental subject whose fate is in critical condition. Such gestures of framing a planetary condition draw from specifically European tropes of epistemological and experiential crisis. Attention to the tropological consistency across realisms discloses how the Anthropocene and new forms of philosophical realism that have been forged in response to climate change are not divergent trends in ecological thought, so much as they give face to two sides of the same medallion.

At this juncture, it may be worth recalling Spivak's admonition of poststructuralist critiques of the sovereign subject, which purportedly diffuse the chimera of unified agency into a multiplicity of subject-effects, but actually inaugurate and conserve a "Subject of the West, or the West as Subject" (Spivak 1988, 271). This inversion—this mirrored reflection that is a structure of self-reflexive identity and is the centerpiece of a longstanding exclusionary strategy of foreclosing the non-western from the space of representation—is precisely what threatens to be sanctioned anew, though now, under the name of environmental conscientiousness. "The human species" is the ontological cover for this operation. And realist traditions provide resources for envisioning such a Subject of knowledge, faced with the prospect of its own dispensability. Noticing how realism's discursive strength derives in part from the (de-personalized) fortitude of enduring the staged absence and fragility of the subject (of the West), we may see how—and here is the second thread—philosophical and aesthetic realisms that reached a tipping point in the early twentieth century were already, quite before their time, discourses of extinction.

Vacant mirror, spectral subject

Let us now consider realism as it is split down the husk—referring, in philosophy, to the rejection of the proposed identity of reality and mind, and in art, to a mode of representation that affirms a mimetic tie to the world "as it is." I would like, further, to hone in on a feature common to both aesthetic and philosophical variants of realism—one that is made manifest, in an intensive way, in Antonio López García's painting "Lavabo y Espejo" ["Sink and Mirror"].[5]

López García played a principal part in the post-war, Franco-era emergence of New Spanish Realism, whose claim to novelty rested on certain anachronistic tendencies, including the practice of working from direct observation at a time when the canvas was increasingly given over to abstraction, or inflected by the photographic image. His gaze is often turned to Madrid, to its untidied interiors (he does not shy away from the bathroom), to un-regaled scenes of the urbane prosaic, and to a cityscape whose untimeliness crosses the sight of the modern and the derelict—not as strangers, but as intimates.

Temperamentally, his work is slow. His methods—meticulous, if not exacting: rulers, a touch of *mathesis,* a rare precision of sense. When painting a tree, he hammers small metal stakes into the earth, so that in shuffling his feet up to these fixed coordinates, he returns daily to the same place, the same perspective. Many of his paintings have taken him decades to complete. A large canvas of Madrid required a short lifetime, 20 years returning to the same street, to revisit the sought-after quality of illumination that fell upon the city for only three hours of the day, during three short months of the year. In 2013, the Queen of Spain ordered him to move into the royal palace and to remain stationed there until he finished a commissioned portrait that he had been working on for 18 years (Govan 2013). The film *El Sol del Mebrillo* keeps company with him, as he attempts to see through his determination to render the sight of light falling on a quince tree in his backyard; as if witnessing a slow-speed chase transpiring between invisibles, we look on as the gradual change in seasons threatens to outpace the patience of his work. The fruit grow faster than he can paint them; they brown before he can brush their ripe color into image. His aesthetic—variously praised or dogged as "neo-academic"—is not foreign to the drive to mastery. But the *telos* of such ambition seems, at times, less oriented toward the domination of content and medium by rational process than the realization of an unreasonable affection for the everyday, whose objects claim his peculiarly monogamous, if not total, attention. Technical exactitude is drawn in the service of conjuring a delicacy of vision.

What I am turning the gaze toward in his work may be first approached by noticing how his vistas of the metropolis feature a remarkable emptiness, reminiscent perhaps less of any other painting than of Eugéne Atget's early photographs of old Paris, as it lay moonlit on the brink of forced disappearance. What Benjamin says of Atget's city of lights might well be said of López Garcia's Madrid: "these pictures look cleared out, like a lodging that has not yet found a new tenant" (Benjamin 2005, 519). If, following Benjamin, Atget turns his urban surroundings into a crime scene, López Garcia slows the bustling arteries of the capital into a still life. Gran Via, the main thoroughfare that was built to signify Madrid's entry into European modernity, appears on the canvas quieted: no cars, no crowds, no pedestrians, no sound. Such spectacles of depopulation—set precisely in those areas expected to be most dense—may arouse the apocalyptic mood or, as Benjamin suggests of Atget, scratch the unsettling underside of those most intimate environments, endowing them with a quality of the surreal that is difficult to shake but also gives rise to a felt remove, an estrangement from the course of things that has the potential to be politically fruitful. In a reversal of the direction of technological development unfolded in the history of art, these paintings, seen from a certain distance, are liable to appear to the modern eye as mechanically or digitally reproduced, but, in important distinction from his photorealist contemporaries, he does not work from photographs. To call these images "hyperreal" is to misrecognize resemblance as simulacrum. Doing so obliterates the untimely quality of his practice, which, when subsumed under such a category, is run roughshod by the concept of a univocal now in which presumably there is no original referent for any copy and subjective experience is afflicted by the delirium of the indistinguishability between reality and stimulation.

Not so with his paintings, whose referents have a life of their own—an endurance that outlasts his ability to represent them, an ephemerality too fugitive to be stilled. What his work discloses is eclipsed just as thoroughly by aggrandizing epithets, the most frequently cited being Robert Hughes's praise of López García as "the greatest realist artist alive" (Hughes 1986, 2). Such claims not only perpetuate the violent mythos of the man-genius as the protagonist of European culture's march forward, casting a long, solitary shadow over the collective of artists—most tenebrously over the women, Isabel Quintanilla, María Moreno, Amalia Avia—who helped to shape the aesthetics of Madrid Realism. In regarding López García as a virtuosic stalwart of tradition that everywhere is threatened by the "tyranny of the neo," or, as an exemplar of an increasingly rare fastidiousness that is the mark of true craft in a degraded present, such criticism not only celebrates a deafness to the understatement that sustains López García's art but also turns him prematurely into a relic (Hughes 1986, 1).

Against the current of the superlative, one may ask, what do his paintings wordlessly speak? Of the works in López García's *oeuvre*, "Lavabo y Espejo" claims the special attention of these reflections, not only because it is a prototypical instance of pictorial realism but also because it offers a visual distillation of the central conceit on which realism depends. The whole surface of the picture is covered in what Barthes would call those "futile details" that produce the reality-effect—extraneous facts, hardly functional elements that add nothing to a narrative trajectory but work to connote *this is real* (Barthes 1989, 141). Notice the faint halo of brown, ringing round the five drainage holes of the sink. The bunched rag at the base of the frame. The subtle veneer of dirt, and how, in certain places, the grout between the tiles grows bolder, elsewhere, lighter. The slight shadow cast by the beaded chain that serpentines its way from the faucet to the drain. All this speaks, in Barthes's words, a "referential plenitude" (Barthes 1989, 148). The painting approaches pure notation or, to switch registers, a pure denotation—as the aggregate of austere objects comes together in a picture of "concrete reality" (Barthes 1989, 146). Thematically, this painting shares with the European realist novel of the nineteenth century a disciplined concentration on the prosaic—on this *oikos*, this home, this closed economy.

FIGURE 5.1 Lavabo y Espejo [Sink and Mirror] (1967); Antonio Lopez Garcia; Oil on Wood; 38 ½ x 33 in.

But what is to be done with the flagrant lapse of fidelity that occupies the center of the field of vision? What to make of this looking-glass which exercises selective vision—capturing a faithful reflection of the perfume bottle, the tip of the shaving brush, but refusing the labor of duplicating the observer? The picture plane does not offer a sighting of the painter who, in order to represent the scene, must have stood before it. Nor, finally, is there any record of the extra-pictorial observer, whose gaze finds nothing in the blankness of the reflection to recognize as her own.

Where has the observing subject gone? Has it been forgotten? Removed? Disappeared? Surgically omitted? Painted clean? The mirror improbably sees no viewer who looks. It forcibly breaks with mimesis and unsees seeing. Its operation is opposite that of the reflective instrument found in the depths of "Las Meninas" (1656)—that baroque image of Velásquez, who is often taken as a point of origin for a (patrilineal) succession of Spanish realist masters, in relation to which critics have elected López García as the most recent heir.

The comparative glance at these two paintings—exemplary of the genre—reveals an historical fissure in realism. In *Les Mots et Les Choses,* Foucault reads "Las Meninas" as an exemplar of a mode of representation that seeks to represent itself. The mirror hanging on the far back wall of the chamber bypasses the represented space altogether and cuts straight through to the alleged exterior of the painting to reflect the image of the sovereign pair that sits outside and in front of the scene depicted. López-García's "Mirror," by contrast, offers only what is already within the field of reference and refuses to acknowledge visually the existence of any purveyor within or beyond this inert world of things.[6] The shallow space remains shallow—foregoing the opportunity to open up the picture plane onto a fictitious depth.

If the mirror in "Las Meninas" draws what is exterior to the painting into its interior—and if the result is an infinite oscillation ("the entire picture is looking out at a scene for which it is itself a scene"), López-García's bathroom generates a convex infinity, propelled by the repeated expulsion of the observer. The gaze of the painter in "Las Meninas" implicates the viewer in the space of representation, draws us into a circuit of visibility by virtue of being seen and made into a scene (Foucault 1994, 14). Velásquez, as Foucault observes, orchestrates sightlines to achieve a remarkable coincidence of three disparate perspectives: that of the painter; that of the figures represented—the queen and king, who look upon the painted scene as they are being painted; and that of the spectator (Foucault 1994, 4). By contrast, López García's painting does the work of subjective extrication. The specular image effects an interruption, halting, and closing off of (the work's) self-reflexivity. To look at the painting is to generate a radical non-coincidence of the subject and its perspective—a cleft whose severity exceeds the split occasioned by psychic or specular disidentification.

Not only does this composition make a spectacle of vacancy, it produces a space that cannot be entered. That is, the space cannot be entered if the referential claim is to be honored. Medium, here, is crucial. As a painting, drawn as it was, from

FIGURE 5.2 Las Meninas, Diego Velázquez. Oil on Canvas. 1656. 10'5" x 9'1".

life—and not captured mechanically—the observer, who paints from this frontal perspective, would have to have been deliberately written out.

This effacement generates a movement that is redoubled for the spectatorial viewer. One steps before the painting, but the very instant that its promise of verisimilitude is accepted, it is broken, for the reflection proves untrue. To take in this vision of the "real" is at the same time to be expelled from its space: one enters, one is evicted, one enters again, and one is driven out once more.

The representation, executed under the sign of realism, works to de-constitute the subject in the visual field. This banishment is not a punctual action but is a continual process, renewed every time the painting is seen. Its grammatical mode would be the gerund; for the observing subject is perpetually being removed by persistently being unfigured.

This is precisely the defining gesture staged by modern philosophical realism.

As shorthand, this operation will be referred to as the phantom subject conceit. It is deployed, to mark just one instance, in Cambridge Realism, which is gently and memorably parodied in Virginia Woolf's novel *To the Lighthouse.* When Lily Briscoe inquires about the topic of the philosopher-father's research, his son Andrew replies: "Subject-object and the nature of reality … {T]hink of a kitchen table when you're not there" (Woolf 1989, 23). In this family dialogue, one hears the echo of Bertrand Russell's *Problems of Philosophy*—published in 15 years earlier in 1912—which begins by posing the following question:

> The problem we have to consider is this: … When we have enumerated all the sense-data which we should naturally regard as connected with the table, have we said all there is to say about the table, or is there still something else—something not a sense-datum, something which persists when we go out of the room?
>
> *(Russell 1912, 19–20)*

Within his refutation of idealism, the real will be defined as that which is mind-independent and able to persist when the hypothetical subject takes leave of a place. Russell's experiment presents, in condensed form, how modern philosophical realism validates itself by producing such occasions of self-absention—test-cases in which the real is deduced by means of imagining the subject's removal. As in López-García's picture, the real is where the subject cannot be seen, where, spatially, it cannot enter; such a Realism at once repels the subject and makes it maximally dispensable. The reality of the object-world is secured but only as it is severed from the subject, who remains from the world removed.

The phantom subject reincarnated

Looking forward, one sees the phantom subject reappear, although newly aggrandized, in contemporary attempts to conceptualize ecological catastrophe (even if only as a limit to conceptuality). Dipesh Chakrabarty's "Four Theses on Climate Change"—to take a paradigmatic instance—places the realist trope at the center of his appraisal of global warming, even though he does not acknowledge it as such. He opens by citing Alan Weisman's *The World Without Us*—a book whose central thought experiment, encapsulated in the title, is said to express the foremost challenge posed by the Anthropocene. Marking the inadequacy of historical materialism for apprehending the deeper histories of the geological Real, Chakrabarty reprises the realist gesture of vacating the subject and installs it as the defining reality-principle of the present era of planetary warming. Says Weisman (quoted by Chakrabarty): "Suppose that the worst has happened. Human extinction is a fait accompli …. Picture a world from which we all suddenly vanished …" (Weisman 2008, 197). Like a time-lapse camera, Weisman narrates with great rapidity the gradual decline of the built environment over the course of hundreds of years. The entirety of the second chapter is a protracted imagining of an unpeopled Manhattan taken under by "water's retaliation"—the Atlantic currents rushing through the subway, the pavement cracked open by weeds, the money in the bank-safes mildewed, art molded, coyotes roaming the wilds of Central Park (Weisman 2008, 31).

The sketched scene is not without flourish, nor without a kind of satisfaction—a cousin, perhaps, of those Romantic lyrics that fantasize about nature having its day, overgrowing the monuments of civilization, while the poet is somehow there to imagine it. The speculation, moreover, hangs on certain presumptions that might be more clearly brought forward as anthropo-narcissistic investments—not least of which is the affirmation that the "worst" imaginable eventuality would be "our" own death. Chakrabarty, nevertheless, takes the "world without us" postulate as emblematic of the "historicist paradox" induced by climate change and the cultural anxieties that attend it: the possibility of a radical discontinuity between the present and a future devoid of humans, and thus "beyond the grasp of historical sensibility," threatens the very continuity of human experience that subtends "history" (Chakrabarty 2009, 197). Historical understanding, he suggests, is "thrown into deep contradiction and confusion"—bereft as it is of resources for visualizing a future in which the human is absented. As we have seen, the aesthetic and philosophical imagination is more experienced in working through such a predicament, discursively and visually. But perhaps even more important is to observe how the culturally specific tropes and image-repertoires of realism are presently being redeployed in discourses of climate change, though unmarked, so as to depict a generalized, if not universalizeable, condition.

Elsewhere, in what is called "speculative" realism, Quentin Meillasoux turns, not to the distant future but to the deep past—a retro-projected anteriority through which he likewise spectralizes the human. He does so in order to refute correlationism—his broad term for any critical philosophy that, following Kant, treats subjectivity and objectivity, thinking and being, as inseparable. By treating the prehuman world as a test-case of such a position, he aims to sever dramatically what exists from any subjective attachment to it.

He suggests that proponents of correlationism are afflicted by a "bereavement" with which they have not come to terms (Meillassoux 2010, 7). This conceptual deprivation is framed both as a claustrophobia ("We are in consciousness or language as in a transparent cage. Everything is outside, yet it is impossible to get out" [6]) and as an ecological loss, which tacitly evokes a melancholic view of a world in which the process of conquest has reached a stage of completion and geographical exhaustion. There remains, in other words, nothing left outside "us" (read: recognized subjects of European philosophy) to appropriate:

> Contemporary philosophers have lost the *great outdoors*, the *absolute* outside of pre-critical thinkers: that outside which was not relative to us, ... existing in itself regardless of *whether we are thinking of it or not*; that outside which thought could explore with the legitimate feeling of being on foreign territory—of being entirely elsewhere.
>
> *(Meillassoux 2010, 7)*

Through a lament about the loss of foreignness, in which the resonances of a postcolonial moment break through the veneer of ahistoricity, Meillassoux intensifies the Russellian criterion of accepting as real only that which is indifferent to human thought. He does so by placing special significance on the "ancestral" realm—a state of reality anterior to the emergence of the human life, and thus to any possible correspondence between subject and the world. This gesture toward geological past is a means of positing what is subjectively unmediated, or absolute. This retrospective subtraction of the human—a rather literalist take on breaking the correlationist circle—amounts to a speculative bid for the *in-itself*, which Kant's humility had set aside from the reach of reason's certainty. Turning away from linguistic mediation, Meillassoux elevates mathematics for its singular capacity to "discourse about a past where both humanity and life are absent" (Meillassoux 2010, 26). A realism, in short, of a world without us. If the opening quotation of this chapter may be reprised, one may see how such a stance coincides with the definition of realism supplied by Piaget via Minh Ha as the act of "ignoring the existence of oneself and, consequently, in taking one's own view for immediately objective and absolute" (Meillassoux 2010, 55). Math becomes the mask that sounds the voice of immediacy, issued from a subject pretending not to be there. Such objectivity is a supreme fiction—a fantasized persistence beyond the limits of the subject, joined with a power play that makes vanish the subject's own relationality to the scene of knowledge or perception.

Speculative realism indexes a cultural moment in which, it is not, as its advocates contend, that epistemology is defunct; it is, rather, that one of its central conceits has been newly amplified through the generalization of the observing subject so as to become the doublet of the human as species. The provocation, in other words, is not only to imagine what happens when a hypothetical witness leaves the room but to consider what remains when humans as a whole are removed from the face of the planet.

Having traced the phantom subject conceit through two of its most prominent contemporary incarnations, I would like to underscore how curious it is that in order to cope with the ontological precarity of the object-world these discourses repeatedly enact the disappearance of the subject of experience. In other words, such realisms secure for the human the staying power of the world it inhabits, but this constancy is underwritten by the fantasy of the subject's absence.

Though realist discourses are often understood to stabilize a picture of the external world, I am suggesting that they are also a cultural means of working through and fantasizing about the human subject's extinction—collective or individual. Recall how in López-García's painting one can see the space of lived experience but cannot quite enter it. It is not incidental that modern forms of realism—distinct from their Platonic or medieval predecessors—begin to be forged within a cultural milieu in which the ecological devastation (*oikos*: house, dwelling) engendered by industrial capitalist modernity was beginning to be registered by the collective sensorium. Although irreducible to a straightforward symptomatic manifestation of economic disturbances, modern realisms in arts and philosophy may be partially understood as a discursive encoding—as genre, aesthetic, standpoint—of a historical condition in which the world remains available to perception but is sensed to be on the verge of becoming structurally uninhabitable, if not unlivable. Whereas the surrealist optic vacates the streets so as to make the fabric of the everyday shudder into a felt estrangement that can be politicized, its realist counterpart formalizes estrangement and arrests it as an image of the objective—"as it is."

Thus, the truth-content of realisms resides less in the referential claims that they make than in the *Spaltung* that has been crystallized within them. As a representational mode, such realisms formalize and render insuperable the division between experience and knowledge. They generate a vision of the world sundered from the subject's ability to imagine herself within it.

Planetary fort-da

This chapter has brought into view the phantom human subject, whose imagined extinction is discursively productive in that it generates for realism the referent of "the real." Apprehending this discursive fixation allows for a clearer vision of how this conceit of realism contains the negative image of the conceptual paradigm of the Anthropocene. Where realism emphasizes the finitude, inadequacy, if not total expendability of the (knowing and perceptual) subject, the rubric of the Anthropocene conjures an omnipresent, transcendental subject, a unified protagonist—*anthropos*—who is newly invested with telluric agency.

The Anthropocene, which Donna Haraway has aptly identified as an "extraordinary abstraction," announces itself as a new geological era defined by the actualization and recognition of humankind's unprecedented power as a geophysical force (Haraway 2014). Though it poses as a trans- or supra-cultural phenomenon, the Anthropocene is the expression of a problematic specific in its frame and formulation to Western modernity, not least because it recapitulates the stagist trajectory of self-awareness that comprise the ideological nerve of Enlightenment philosophies of history. The Anthropocene is the name for a moment of self-consciousness, when what was practically set into motion by industrialization is belatedly acknowledged, consolidated into a proper name, and raised to the status of a geo-historical epoch, which is then positioned as the culminating point of a linear, developmental continuum, over which the specter of catastrophe looms.[7]

Rhetorics of the Anthropocene reflect less an act of coherent periodization than a reprisal of an old ambition—though appearing under a novel, scientific signifier—to engineer a monumental discourse of *Weltgeschichte,* in which the (European) human becomes the author of universal destiny—even if, in this round, it may very well culminate in demise. At this late moment, skepticism toward progress is not sufficient to undo its spell. The "human" figured by the Anthropocene resurrects a transcendental subject at the very moment that such an enterprise, theoretically speaking, became moribund, if not indefensible. In the name of environmental awareness, such a concept reinscribes the coherence of the *anthropos* and once more imprints with the insignia of scientific validity the discursive consolidation of the human—a spiritual, abstract ideal of European projects of civilization, which, since its inception, doubled as a technique of racialization. The centrality of the Western subject is thus preserved, admittedly in a more anxious formation, but perhaps more enduringly, since the *anthropos* is shown to survive the prospect of its own negation.

Understood differently as a hermeneutic, the Anthropocene redescribes phenomena of climate change in terms of a crisis of management—management of the unintended consequences of "our" action, which include the imminent possibility of self-annihilation. Rhetorics of the Anthropocene train one to see behind every natural disaster the invisible hand of human action, denuded of its geopolitical specificity. While such a panoramic narrative does offer a way to mark the unprecedented scale and degree of anthropogenic intervention into the self-regulating processes of the planet, it does so at the cost of enacting epistemic violence—since it establishes as the dominant theoretical framework the problem of preventable self-destruction, when more often, as the fragments of archive show, it is more accurately identified as selective homicide.

Its bid for status as a general theory or as a scientifically authorized period of history is patent; less explicit is how any time the Anthropocene is invoked an ontological claim is being made. As a conceptual frame it ontologizes the (Western) opposition between nature and the human—and arranges that relation into a pattern of self-reflexivity. This harbinger of a geo-historical change in fact works to hypostasize and render timeless the historically contingent cleft between humanity and everything else. This fault line is the very same that, in realism, divides the phantom subject from the matter of the world.

In the humanities, two prevailing theoretical tendencies for addressing climate change in the academy of the global north—namely, the analytic of the Anthropocene and the revival of realism and ontology—must be treated, not in isolation but as each other's doublet. Both rooted in Eurocentric intellectual traditions of representing reality and fantasizing about one's own death, precipitated by reflexive circuits of productive activity—they generate inverse modes of perception. One can visualize this contrast in terms of obverse specular dynamics. Velasquez alongside López-García: the mirror that makes the depicted world into closed and thus infinite circuit for reflecting the human sovereign back to itself; and the mirror that shows a haunting vacancy, offering the peculiar satisfactions and pains of making perceptible to the human the scene of its own eradication. The Anthropocene encourages us to see the human and its undertakings everywhere we look, and the realist gaze renders a vision of an everywhere, in which the human is nowhere to be seen. Repressed subjectivism converges with the planetary narcissism that is its conceptual opposite.

It is hoped that climate realism may be navigated theoretically without having to submit to the potentially monolithic structure of these false alternatives, or put otherwise, to the deranged game of fort/da on a planetary scale, which shuttles between extreme presence and absence as the only modalities for encountering loss: an oscillation between, on one side, the fiction of human powerlessness that is so severe that it can only imagine reality by sentencing the human to eradication, and on the other, the hyper-inflated agency of the universalized protagonist of the Anthropocene. These two optics are bound to each other as dialectical antipodes but have a special kinship in the way they refurbish older discourses by making a leap of substitution, from subject to species. Without the abstraction of *species* as an historical invariant, these discourses would come unfurled. In this stark strobing between the all and the nothing—the *anthropos* that is too much with us and too far gone—one catches a glimpse of the emancipatory otherwise, in which it is not only *man* that, like a visage drawn in the sand is effaced by wave, but *species* that redeems the oppressed past at the moment when luminous, it at last gets gone.

Notes

1 Raymond Williams charts, with characteristic clarity, the transformations in "Realism," beginning with the "Realist" opposition to nominalism. This chapter is primarily concerned with modern realism, whose emergence Williams dates to the nineteenth century (1830s in French, 1850s in English). Earlier senses of the "Real," as he points out, pertain to law and property (in reference to things that actually exist and can be sold); or refer to that which underlies appearances. In addition to the now less common sense of realism as a philosophical belief in the existence of (Platonic) universals, Williams mentions three other definitions, which will be placed in circulation: (1) "a term to describe new doctrines of the physical world as independent of mind or spirit"; (2) "a description effacing up to things as they really are, and not as we imagine or would like them to be"; and (3) "a term to describe a method or an attitude in art and literature—at first an exceptional accuracy of representation, later a commitment to describing real events and showing things as they actually exist" (258–259).

2 Seen in this light, the literary genre of psychological realism is less revealing as a technique of classification than as a trace of the remarkable intelligence with which writers like Woolf resisted the sleight of hand that would pass off a geopolitically contingent reality as something that holds universally and instead attempted to countenance the complexity of truth at that very ridge, where the risk of contradiction warbles in all its terrible fullness.

3 The putative power of realist discourses recalls Eve Sedgwick's insight into the paranoid timbre of forms of theory that Silvan Tompkins identifies as "strong." The stronger the theory, he maintains, the greater its ambition to enlarge its domain to account for phenomena that are remote from one another (Sedgwick 2003, 134).

4 For an account of Foucault's understanding of the "realism" generated by biopolitical governance and the statistical sciences that make it possible, see Chapter 7 by Ingrid Diran and Antoine Traisnel in this volume.

5 My thanks to Ana Schwartz who one day returned from the library with a monograph of Antonio López-García's work, which she shared with me in the eye of winter.

6 For those interested, one can listen in on a conversation that López García and a friend have about "Las Meninas," which is included in one of the discarded scenes of *El Sol del Membrillo*.

7 Both Chakrabarty and Donna Haraway remark upon the historical conditions that enabled the Anthropocene to become legible as an object of cognitive apprehension. Climate science, he reminds us, emerged in the context of militarized research in space exploration, which sought to determine, among other things, whether Mars would be habitable as a base for fighting space battles. Haraway claims that the technological preconditions of an Anthropocene optic included post-war developments in systems theory and cybernetics—which made it possible to process large amounts of data—as well as the newfound access to a literal vision of the planet, seen from outer space.

References

Barthes, Roland. *The Rustle of Language*. Translated by Richard Howard. Berkeley: University of California Press, 1989.

Benjamin, Walter. "Little History of Photography." *In Selected Writings: 1931–1934*. Translated by Rodney Livingstone and others. Edited by Michael W. Jennings. Cambridge: Harvard University Press, 2005, pp. 507–530.

Chakrabarty, Dipesh. "The Climate of History: Four Theses." *Critical Inquiry* vol. 35, no. 2 (Winter 2009), pp. 197–222.

Foucault, Michel. *The Order of Things: An Archaeology of the Human Sciences*. New York: Vintage, 1994.

Govan, Fiona. "Artist Told to Finish Painting of Spanish Royal Family He Started 18 Years Ago." May 9, 2013. *The Telegraph*. www.telegraph.co.uk/news/worldnews/europe/spain/10046798/Artist-told-to-finish-painting-of-Spanish-royal-family-he-started-18-years-ago.html. Accessed May 15, 2013.

Haraway, Donna. "Anthropocene, Capitalocene, Chtulucene: Staying with the Trouble." Lecture. In *Anthropocene: Arts of Living on a Damaged Planet*. May 2014. http://opentranscripts.org/transcript/anthropocene-capitalocene-chthulucene/.

Hughes, Robert. "Art: The Truth in the Details." *Time*. April 21, 1986. http://content.time.com/time/subscriber/article/0,33009,961176,00.html. Accessed November 6, 2017.

Jameson, Fredric. *The Political Unconscious: Narrative as a Socially Symbolic Act*. Ithaca: Cornell University Press, 1982.

_____. *Postmodernism, Or, The Cultural Logic of Late Capitalism*. Durham: Duke University Press, 1992.

Lukács, Georg. "The Ideology of Modernism." In *The Meaning of Contemporary Realism*. Translated by John Mander and Necke Mander. London: Merlin Press, 1963, pp. 17–46.

Meillassoux, Quentin. *After Finitude: An Essay on the Necessity of Contingency*. Translated by Ray Brassier. London: Bloomsbury Academic, 2010.

Minh-ha, Trinh T. *Woman, Native, Other: Writing Postcoloniality and Feminism*. Bloomington: Indiana University Press, 1989.

Russell, Bertrand. *Problems of Philosophy*. Indianapolis: Hackett, 1912.

Sedgwick, Eve Kosofsky. *Touching Feeling: Affect, Pedagogy, Performativity*. Durham: Duke University Press Books, 2003.

Spivak, Gayatri. "Can The Subaltern Speak?" In Nelson, Cary, and Lawrence Grossberg, eds. *Marxism and the Interpretation of Culture*. Urbana: University of Illinois Press, 1988, pp. 271–314.

Williams, Raymond. 1973. *Keywords*. New York: Oxford University Press.

Weisman, Alan. *The World Without Us*. New York: Picador, 2008.

Woolf, Virginia. *To the Lighthouse*. New York: Harcourt, 1989.

6

GEOLOGIC REALISM

On the beach of geologic time

Kathryn Yusoff

You Are Here

Prologue: *you are here.*

Let's start with the end of the World, why don't we? Get it over with, and move onto more interesting things.

—*N. K. Jemisin, The Fifth Season*

The end of the world is the nothing new that N. K. Jemisin's award-winning Broken Earth trilogy begins with.[1] Broken Earths have a long history—colonial invasions and subjugations, the seismic rifts of Middle Passage continental dispossession and indigenous extermination, the reorganization of island climates, and the introduction of "alien" ecologies. While the Anthropocene narrativizes a phase shift in planetary relation—the climate pause is off, geologic realism is on—the intersection of geography, colonial power, and its material space-times establishes an imaginary of arrival in environmental crisis that normalizes a trajectory of colonial power and its subjective forms. As the Anthropocene asks us to look at the mineralogical subsurface beneath our feet, and anthropogenic climate change bids us to look at the portending sky, one offers a mutable heaven of weathering to come, and the other tells of the banal realisms of extinction within the abysmal sands of geologic time (and its ongoing weathering). Looking both ways along the axis of sky and ground, something has shifted in the way thought and materiality are tied together; something is out of joint and materially unbound, between a rock and a hard place in an account of geologic time. Late liberalism is now reckoning with the possible end of its world, in which the Anthropocene is presented teleologically, as the end of late liberal modalities of life. Meanwhile, other worlds have long since been surviving their enforced apocalypses of subjective deformations and ecological deaths raised in the context of colonialism and slavery and in the ongoing presents of settler colonialism and antiblackness.

This leads to the question of how we theorize geologic realism and the "end of the world," given the foundational entwinement of colonial categorizations of inhuman matter (a geophysics of extraction) and inhuman subjection in the shadow of the Kantian subject (a metaphysics of race). As global colonization by Western powers proceeded by way of extraction and subjugation, in relation to indigenous and black subjects, what nonarrivals, extinctions, and erasures need to be considered in the locative and universal command of "you are here?" As Jemisin's novel indicates, finding "a way out of no way" has long been the work of survival in the rift of racialized dehumanizing worlds. To wit, the space of the inhuman, between a rock and a hard place, is a historically occupied (non)subjective position.

While the Anthropocene is a new geologic time interval that pries open a speculative dimension to planetary thought and material relations, it also requires that the histories of matter and their subjective-racial formations not be forgotten. In the rush to a postracial humanity, understood as a planetary geologic force, the geologies of race that constituted the foundation inscription of the discipline of geology and its praxis in colonial world making are being erased. In this confrontation with geologic realism in the Anthropocene, the end of the world is up for grabs both in the acknowledgment of its past iterations and in the configuration of its futurism (from reproductive futurism of heteropatriarchy to the future perfect tense of black feminist thought). The very context in which thought exists and takes hold of matter is being put up for question as an environmental rather exclusively theoretical event, both in terms of the (colonial) "time" of thought and its occasioning in the "happy Holocene" of climate stability, and in terms of how thought forms its (subjective) objects through its apprehension and classification of materiality. An example is the figure of the indigene, whom Elizabeth A. Povinelli argues has long been positioned as a living fossil in the politics of late liberal recognition (2016, 77) who is further sedimented outside of the political present in the valorization of the Holocene as the desired planetary state, placed either in romanticized pre-Holocene paleo-personhood or as a subjective resource for configuring a post-Holocene survivalism or a narrative of pernicious resilience. As with seventeenth-century geologies of race that used a combination of fossil objects and anatomy of "racial types" to hierarchize the human into its categories of Man and his others, the temporal inscription of matter makes political claims on the past and present.

Geology is an operative praxis of materiality and thought in the genealogy of Western metaphysics and racial capitalism and a renewed site of possession in the Anthropocene, whether that geologic matrix is proffered as a "saving" of the conditions of late liberalism or through the continued racialized configurations of exposures to environmental extraction and impacts. Thinking with the relation between thought and matter is a way to raise questions of how futurity gets shaped through temporal lines and how imaginaries of geologic time consolidate particular subjective and nonsubjective (or inhuman) placings. I want to think with the visibility of these subject–object positions of inhuman matter in the context of a reinvestment in colonial mythmaking about human master subjects in relation to the geologic forces now unleashed and the modes of valuation that inhabit that propagation. Another way to pose this would be to question how investments in certain formations of the Anthropocene usher in obligations (or not) to reproduce the current colonial-matter-matrix and its extractive relation to the world, and how imaginaries that refuse this extractive praxis and its subjugating conditions can force the recognition of different materializations of the world (or world making).

I want to use the analytic of geologic realism to navigate between the representational ethics of a geologic kitchen-sink realism (a fuller Anthropocene told from the messy raced, sexed, gendered, and economic relations of a geologic undercommons) and the material propositions of speculative realisms (in the shadow of Kantian subjective claims on materiality and knowledge, which organize toward a mastery of matter).[2] I understand geologic realism to be a way of thinking through the entwinement and emergence of (a) normative modes of approaching matter (forged through the histories and temporalities of colonial power in material and metaphysical orders); (b) the realism of quotidian experience by black and indigenous peoples in the shadow of the imposition of colonial power; and (c) a shared political present of planetary shifts and geologic disruption (related to colonial configurations of matter). These juxtapositions of subjectivity and material relation are discussed on the beach as a contact zone between apprehension and erosion on the littoral/lithic shores of time.

The narrative of encounters on the beach is staged as analogous to three distinct Anthropocenic subject positions in relation to the inhuman: humanism (*Planet of the Apes*), speculative realism ("The Terminal Beach"), and black feminist futurity (*Daughters of the Dust*). In a promiscuous interdisciplinarity, the vignettes speak to the desires and possibilities of inhabitation that are interned in temporal lines, where the beach acts as a space of reckoning with relation to time and materiality and puts pressure on those sightlines and their horizons of meaning. In this analysis, I prize out some of the ways in which the temporal understanding of materiality is a means to signify and subjugate agency in the political present in the breach of thought and world. Although different in genre, the narratives are considered together, in relation, for how their imaginaries of subjectivity and geologies of time might be made to crash against one another and their aesthetics touch different registers of temporality and relation in world making. The topic of the last vignette, *Daughters of the Dust*, offers a way of seeing agility in relation to temporal shifts with respect to these prior formations of racialized patriarchal subjecthood, and one that radically pushes for remaking the terms of survival and subjectivity in a different conjunctive mode of materiality and time.

These differently located encounters are staged in what Édouard Glissant calls "a going-on-with the dialectic among aesthetics," acknowledging the relation between imaginaries and temporal plurality at play in the meeting of New/Old Worlds (1990, 203). In Glissant's thought, the idea of archipelagos is a way to practice a located totality against the attempted monolinguism of colonial framing of Earth as territory (as a precept for extraction). Similarly, the beach might be a place to encounter the lithic, liminal, and littoral scene of geologic realisms that wash up in the tidal zones of materiality and metaphor. As a space of geologic weathering in the materiality of time, the engagement with the beach is made in the context of a history of engaging the fluvial shores as an analytical space in black geographies, from the archipelago poets Aimé Césaire, Glissant, and Dionne Brand to the black feminist "wake work" of Christina Sharpe, Katherine McKittrick, Tiffany King, Michelle Wright, and Vanessa Agard-Jones that refutes "antiblackness as total climate" (Sharpe 2016, 21). In her work on shoals, King argues littorality regarding "relations, aesthetics and associations that are 'off the shores' of normative orders of knowledge. The shoal is place where the sea and non-sea matter (rock, sand bed, coral) meet. As both sea and non-sea, or neither sea nor not-sea, it is an ecotonal space where different kinds of ecological systems meet" (2018, 21).

I am interested in the role of geology in temporalizing subject positions to maintain racialized geosocial formations in the emerging context of a consideration of what might be called black Anthropocenes. Broadly, these are engagements with the Anthropocene that insist on a politics of recognition of the apocalypses of black and brown subjects within the ecocide of colonialism and empire. It is a politics of recognition that foregrounds the visibility of another inhuman history within the historicity of the ongoing coloniality of thought and relation in matter economies. I take up these questions of weathering through an engagement with geologic realism to understand the context of the mythic and epochal claims of geology and of geologic force as a newly designated quality of subjectivity in the Anthropocene, a subjectivity that oft forgets its geographical passages. In the context of the new and mythic, I want to stake a historical claim on a missing black and indigenous geologic archive.

The matter of thought

The question of thought's materiality needs to be posed in relation to the genealogy of Western thought as much for what it leaves out of its accounts of the World as for the way Earth becomes within these trajectories. Considered either through the placement of Earth as an existential outside to Western thought or through the stabilizing of dynamic grounds on which the architectures of reason are built, the earth/ground became a point of negation for Western reason even as it spawned whole edifices of meaning to cover over that suppression of the real. Much like Kant's overcompensation for the geologic nihilism of the Lisbon earthquake (1755) in his formation of the sublime (and concomitant racial/racist theories), the material exclusions of geologic agency are as much a product of psychic terror as they are of the geotrauma of earth events. That is, world building is a metaphysical accomplishment as much as it is a geophysical mastery of material registers. Geology represents a special kind of ground and grounding imbued with geophysical force or gravity. It seems geology in Western reason can be thought only through death, where death of subjective thought is tied to catastrophic death in the end of worlds, conjoining the subject and world in a phenomenological bind that refuses the world's ability to go on before and after that thought (and subject). Western philosophy since Kant aimed at containing the realism of the world's continuance after extinction within a representation that adheres to a notion of human meaning and purposefulness on Earth rather than through the unbounded openness of its geologic relation to the cosmos. The tradition of natural philosophy sought to overcome or transcend the universalism of extinction events (as they were being inscribed through the fossil facts of paleontology) with Cartesian logic and rationalism (defined most extensively by Kant and Hegel). It was thought forged in the materialization of a geologic praxis, in colonialism and slavery (1492 onward), and in the formal development of the discipline of geology (1700s).

Geology delivered two kinds of realism: a realism of racial substrata labeled *inhuman* that comprised the subjective resource for extraction (and resistance), and a realism that was an epistemic-ontological confrontation with the history of the dynamic Earth events, told as a tale of beginnings and endings, of catastrophe and survival, of Beings in time.[3] To say that geologic realism outside of the narrative of origins and endings has been problematic for Western thought is an understatement. Only through the retethering of geologic origination as a supplementary genealogy that amplifies human exceptionalism has geology been psychically brought inside the human story, told as the epic survival story of *Homo sapiens* where psychic and material location is plotted out to confirm the sanctity of here and now for some and "you are not here" for racialized others. *Now* is being told as a form of anthropogenesis of geologic forces through a postracial *we* that obscures the historical contours of race in the genealogies of species life (Yusoff and Thomas 2018, 52).

In Western natural philosophy, Earth has successively been pushed away as a problem for thought or excluded to a constitutive existence outside of or anterior to reason. Earth is materially positioned with reason's "others" (the inhuman, the subhuman, and the less than human) in its exclusion from the center of humanist thought, while it serves the realization of that production as subtending strata (as context, resource, buffer for earth shocks, bounty, reproduction, and labor) and as an exclusionary inclusion of matter without the individuate identity of the Western ethical subject (as racialized [non]subject). The earth and the slave as concomitant categories have no recognized identity in the extraction economy of geologic grammars outside their valuation in categories of matter. Geologic realism, then, offers a speculative opportunity in the engagement with geology not just as a means to unearth an anterior posthuman or inhuman position beyond humanism and its implicit reproduction of white heteropatriarchy but as a perspective that could come to terms with both the cosmic potentialities and vicious subjectifications of geologic relations (in racialized, gendered, and sexualized forms).[4]

One corollary of geologic realism is to take account of the asymmetries of inhuman nature (Clark 2010). As Ray Brassier suggests, there is an "unavoidable corollary of the realist conviction that there is a mind-independent reality," (2007, xi) an asymmetrical reality that is oblivious to humanity and its concerns, a reality that cannot be made over as our home or ground, or worked into any kind of "meaningful" relationship to us. The other corollary is the recognition of the constitutive hinge that holds humanism and its production of hierarchical subjectivities through racialized "scenes of subjection" together with a production of the earth, both categorized as inhuman (Hartman 1997). The birth of the racial subject is tied to the material categorizations of colonialism, through the desire for gold, the conquest of space, and the codification of geology together with indigenous and black personhood as a resource praxis. Geologic resources and bodily resources (the extermination of indigenous peoples and commodification of racialized slavery) share a natal moment in the dual exploitation of subjects and Earth through the geologic grammars of the inhuman. As Achille Mbembe comments,

> The question of the world—what it is, what the relationship is between its various parts, what the extent of its resources is and to whom they belong, how to live in it, … what its borders and limits, and its possible end, are—has been within us since a human being of bone, flesh, and spirit made its first appearance under the sign of the Black Man, as *human-merchandise*, *human metal*, and *human-money*. Fundamentally, it was always *our* question. And it will stay that way as long as speaking the world is the same as declaring humanity, and vice versa.
>
> *(2017, 179)*[5]

Riffing off of Denise Ferreira da Silva's notion of racial knowledge and power as the construction site of global space, in which *whiteness* is the color of both universality and geography, the formation of the earth as world-object through the conquest of the New World could be understood as a praxis of global-world-space that establishes a world (white)/earth (black and brown) bifurcation. The afterlife of this racialized materiality in the Anthropocene rebounds with questions about the double life of the inhuman, as both inhuman geologic matter and inhumane racialization of personhood coded as matter. These racialized materialities are the constitutive outsides/exclusionary inclusion of humanism, as inhuman (Earth) and inhuman (race), and the hinge between them depends on "*racial subsidies* to exploit the planet's resources" (Mbembe 2017, 179). There is a double extraction: race is materialized via inhuman matter (slave as chattel, and gold) and (non) personhood (labor, flesh, and fungibility). In the reification of Earth in formation with subjective modes outside of white Western Man (deftly articulated by Sylvia Wynter), racial subsidies are what form late liberalism's substratum or extraction zone, thereby demanding a radical revaluation of what Césaire called the "measure of the world." The historic assault on blackness made in proximity to the codification and valuation of the inhuman (as earth and race) established intimacy with the inhuman and made it a site of radical revaluation in critical black feminist thought and materialities.

However, a more complicated understanding of the subjective life of the human does not mean giving up on the radical alterity of inhuman matter; rather it might be seen simultaneously as alterity and intimate possession. As Nigel Clark has argued, the inhuman does important work in situating subjectivity in a planet in which processes and forces are (in) differentially shared across human, nonhuman, and inhuman entities and their temporalities, rather than walled up in the individuation of impoverished versions of subjectivity that externalize that relation into neoliberal economies of environmental valuation.[17] In the historical grammar of geology, the natality of the inhuman—as mineralogical and antiblack—is tied in a liberatory pursuit against the effects of extraction and the racial calculus of valuation. Before looking at how imaginaries of temporality produce racial subsidies alongside a conception of the earth, I want to outline the two subjective positions of the inhuman as earth and race.

Inhuman (Earth)/inhuman (race)

The problem of the earth, as it were, is a problem of ancestrality and the multiple originations of Earth(s). That is, Anthropocene science is articulating that there is not one but many Earths, preexistent and possible, within this particular geochemical-cosmic milieu. The significance of this material multiplicity, which has come to the fore with anthropogenic impacts on Earth processes, has disrupted the idea of planetary homogeneity that reciprocates or grounds the notion of universalism (which established the domains of reason and biocentrism and later Gaia or one-worldism). Reading fossil traces of previous worlds as fragments rather than as universalizing explanations, geology has the potential to disrupt and fracture the temporal totality that seventeenth- and eighteenth-century geologists and geographers sought to establish. If Kant's geotrauma ended in a psychic salve of reason and racism to stem the fear of an inhuman and black planet, then speculative realism engages Quentin Meillassoux's concept of the arche-fossil as an "ancestral phenomenon if not a paradigmatic case of a thing-in-itself" (2008, 3) to address an indifferent earth.[6] Brassier asks: "The arche-fossil harbours a transcendental enigma about the possibility of rehabilitating the thing-in-itself. The transcendental question provoked by the ancestral phenomenon to speculative realism is: How is thought able to know an object whose existence does not depend upon some constituting relation to thought?" To resolve this, Brassier suggests that there is a need "to be able think terms independent of the primacy of their relation, and hence to think the primacy of the object over any of its relations to things, whether they be thinking or nonthinking" (2007, 64). Such a cut in relationality, or asymmetry, is what is understood to be delivered up on the fossil shore; fossils are the leftover forms of organisms and creatures that no longer have belonging in the world, like the light from stars, they make impossible temporal journeys of inscription and erasure. In their captured state, fossils attest to the improbable arrest of dying in the right geomorphic place and leaving a good fossil trace, but they also attest to the inverse of this, which is the outside of genealogy, the leaving of no trace, the absence of a recognizable being in time, a nonhistory, a missing archive (like the black archive).

Inverse to geologic accumulation of fossils as a form of sedimentation (as it is told in the natural history museum), Glissant speaks back to the realism of nonhistory as a subjective condition "through the accumulation of sediments" (1990, 33). Placed on a different terminal beach from Brassier's, he reconstitutes silt as a residue of descendants made in the Middle Passage–plantation matrix and the Piton eruption, all of which has duration, giving life to new forms and relations: "The tie between beach and island, which allows us to take off like *marrons*, far from the tourist spots, is thus tied into the dis-appearance—a dis-appearing—in which the depths of the volcano circulate" (206). On another alluvial shore, Mbembe describes how "the world that emerged from the cannibal structures is built on countless human bones buried under the ocean, bones that little by little transformed themselves into skeletons and endowed themselves with flesh." He continues: "The durability of the world depends on our capacity to reanimate beings and things that seem lifeless—the dead man, turned to dust by the desiccated economy" (2017, 181). In other Earths contemporaneous with this one, those that are violently impacted but without resource to the privileged of Western reason (and inscription in its archive), the possibility of the world relies on a recuperation and animation of those that do not have a consecrated archive, that live in what might be understood in Fred Moten and Stefano Harney's terms as a racialized material "undercommons," of geology and its nonrepresentation (2013, 1).

In one account of geologic realism, the fossil is a chance encounter, a constellation of a very particular set of material circumstances that organize the signifiers of geologic life, materially transmuting the dead into enduring effigies beyond the moment of their Earth and their temporal occasion within it. This mineralogical muscularity, of anteriority, summons ancestors and genealogy to petition on the historicity of forgetting. In this sense, fossils become more and less than their constituting relations, severed from all the conditions and interactions that have made them in time and abandoned on the shore into another time of uncovering. That the rapid accumulation of fossil knowledge emerged through colonial extraction regimes is not considered part of this ancestry. Yet, the existence of fossil objects that have meaning relies on a whole speculative armature of world making that conjures their presence through incomplete material residues. Fossils, then, are understood as empirical knots of time that are without corresponding experience, which operate at the level of classification and speculative fabulation. These temporal objects both sediment extinction in realist terms and push us to the thought of extinction, a thought that can never quite be fully accommodated in the same realist terms within the contours of Western thought. That is, Western reason cannot imagine its own extinction as a system of thought without the end not just of its idealized subject but of the world, and indeed, the two are tied as colonial twins of symbolic and actual world making.

In another account of geologic realism, in the thingness of chattel slavery and its aftermath, extinction and extinguishment were a quotidian realism of the violent subjectification of what Frantz Fanon called the "zone of nonbeing," a realism from which critical black thought emerged as an aesthetics charged with making good on the "reservoirs of life" in this other material margin (1952 [2008], 7). Without recourse to global-world-space or the genealogy of Western man, such thought simultaneously engaged other Earths and relations, modes of innovating from inside the limits of Western reason, toward a differently configured alterity as radical black praxis.[7] As Michael Dash comments about Glissant's work, this involves an exploration of time that is related to neither "a schematic chronology nor a nostalgic lament. It leads to the identification of a *painful notion of time* and its full projection forward in the future, without the help of those *plateaus in time* from which the West has benefitted, without the help of that collective destiny that is the primary value of an ancestral cultural heartland" (1990, xxxv). Glissant's work points to a temporality that overcomes the attempted extinction of black life through materialist and metaphysic relation, a temporality that is central to the memory work of black futurity. Refusing the crushing violence of Western reason, black aesthetic life transmuted the trauma of extinction, equaling that realism with speculative vibrations and material innovations that refuse the binds of what was constituted as (racist) life through humanism, that is, life that is not configured in the same genealogical traditions and forms that rely on a division between the organic and inorganic (life/nonlife) for an account of thought and relation. As Glissant makes clear, "I build my language with rocks" (1990, xi).

Vitalists, according to Brassier, try to escape the "levelling power of extinction" by asserting that the final cessation of physical existence is not an obstacle to the "continuing evolution of life." The horror, Brassier contends, of the "traumatic scission between organic and inorganic" cannot be contained within the psychic economy but nonetheless drives the demand for the veracity of organic life, as an awareness of the inorganic bind that both proceeds and follows life, the particular nothingness of nothing to which the organism will return (2007, 238). "But this 'nothingness' cannot be retrojected into the past or projected into the future; the only temporality commensurate with it is that of the 'anterior posteriority' proper to physical death as that which seizes organic temporality, but which cannot be seized by it" (236). A geotrauma runs through Western thought that drives a repetition of this expulsion of geologic or inorganic time, a temporal trauma that cannot be calibrated within the life of the organism or of the earth understood as a singular global-world-space. While the unconscious experience of Earth and its long history imbues every pore of the organic, the resolution and mineral resolve of Earth eludes being admitted to the formations of conscious experience, an unmanageable extinction indelible on the fossil shore that marks the possibility of thought but remains unmanageable within it (Brassier 2007, 237). Freud understands these compulsive episodes as instinctual in character and suggests that the compulsion to repeat points to the nature of the "drive," as organic life's urge "to restore an earlier state of things which the living entity has been obliged to abandon under the pressure of external disturbing forces; that is, a kind of organic elasticity, or, to put it another way, the expression of the inertia inherent in organic life." "Organic striving," Freud suggests, is toward "an initial state from which the living entity has at one time or another departed," and so the organism is compelled to become inorganic once again, "and we shall become compelled to say that '*the aim of all life is death*' and looking backwards, that '*inanimate things existed before living ones*'" (1991, 309–311).

Thus, the instinct "is the primordial pull back towards the inorganic ... understood as the zero-degree of contraction, or *decontraction*" (Brassier 2007, 235). Such decontraction is not mere telos or return in a conventional sense but a compulsion toward an existence before existence, to the zero of inorganic ground conditions, but crucially made from the positionality of life. Brassier suggests that Freud maintains the "realist thesis" according to which the inorganic precedes the organic and uses this to underwrite the death drive (235). Thus, the inorganic, Brassier argues, cannot be understood simply as a past or future state of the organic toward which life strives (i.e., as an axial condition of life) but must be understood as an "originary *purposelessness*, which compels all purposefulness, whether organic or psychological" (236). The division between the interior and exterior of organic and inorganic life is won at the cost of a separation or shielding from the inorganic outside, a geotrauma of organic individuation that cannot simply be contained by death but harbors an exorbitant potential of the inorganic reaches of the cosmos. Geology gives us that scission, the cosmic excess of an anterior geologic life that is caught temporarily in the realism of life but is not bound by its puny origins or endings.

There exists an isomorphic relation between life and its definition in relation to extinction and its demand to recognize a priori mineralogy. This asymmetry has a double edge, captured in the chance fossils that originated in soft muddy shores, fossils that both destroy and stratify the possibility of meaning in time as a constellation in the substrate of all thought. The geologic realism that greets us on the beach, on the Jurassic coast, for instance, is an axiological predicament, a cosmic annihilation that is "not for us," devoid of the warm rub of vitalism against the inhuman waves and the lithic rubbing of the shore. This confrontation with the lithic processes of Earth is a confrontation with (geo)powers that organize beyond life and without interest in its sense of progress or purposefulness. It is only the distinction made between the inorganic and organic that allows the very concept of "progress" to emerge, but at the cost of the suppression of the inhuman earth and its lack of purpose. And it is at this site of division between the organic and inorganic that racial subjugation is constructed as an ontological horror that campaigns on the senses in psychic and planetary terms, coupling fear of a black and fear of an inhuman planet through the category of the inhuman.

The provocation of the Anthropocene demands an engagement with both these geologic realisms, a confrontation with the abysmal horizons of the inhuman beyond the immediate suture of false ends to these processes and the dereliction of black personhood in the psychic category of the inhuman and the ongoing "experience of the abyss" in "a debasement more eternal than apocalypse" (Mbembe 2017, 7). Whereas Brassier locates the psychic schism in the division between the organic and inorganic, it might be historically located in the annihilation of subjectivity under slavery as a lived condition of brutalized refusal to recognize a right to life or organic striving.

Thought now must occur along the sightlines of a shifting planetary *and* racialized axis in the (re)constitution of the inhuman. Anthropocenic geology appears anew as a future-orientated practice that prompts concerns for fashioning alternative worlds and countermodes of fossilization, questioning the hold of materiality within the organization of life and prompting reflection on how arrangements of life and nonlife might be partitioned differently. Alongside sits a silent archive of fossilized and future lives that have been subjugated under the category of the inhuman. This is the black and brown archive.

While the Anthropocene is problematic in all its assumptions about agency and (white) supremacy of matter, it does signal a shared threshold moment, namely, the demise of the stable material conditions of the "Holocene humanism" that provided the context for the dominance of Western thought and reason. That is, the Anthropocene introduces instability into both thought and material worlds, or the materialization of subject formations through an extractive relation to the earth. That this reckoning is made in the context of a threat to late liberal subjectivity (and its sedimentations of whiteness) is not unproblematic, but the sinewy bonds of geology have been and are made differently. There is a proximity to the grammars of geology, the intimacies of the inhuman as a classificatory and subjective mode, and to the force of geologic events that demands that a counterpoetics reorganize the terms of geologic realism.

In what follows I address three lithic stories of geology that pull out dominant archetypes of subjective positions in the context of the inhuman. The site of these geosocial lithologies is the beach—an inversion in time and space, a wall of geologic realism where time's encoding is washed up, eroded, and uncovered (like Anthropocenic materialities). At the beach the seemingly robust reality of social construction begins to slip, denoting a littoral zone of planetary counterplay among the altered physical environment, the psychic space of temporal location, and its historicity. The terminal beach of geology *is* this slip. As Glissant comments, "The edge of the sea thus represents the alternation (but one that is illegible) between order and chaos. The established municipalities do their best to manage this constant movement between threatening excess and dreamy fragility" (1990, 122). Race, gender, and sexuality work through these imaginaries: in the first two scenes it is the conventional white heteropatriarchal subject that falls through the changed landscapes of time as the guardian of temporal breaches/beaches; on the final shore a markedly different account of myth, subjectivity, and ancestrality marks the sinews of the time of black feminist futurity.

What follows is a recursive story in three parts on the beach of geologic realism.

Planet of the Apes (1968)

The human is in a space suit on the run from geologic realism. The release from the hold of time is the conquest of space. Astronaut George Taylor (Charlton Heston) has crash-landed his spacecraft *Icarus* onto the *Planet of the Apes*, where the conceit of human/white supremacy (in the racially coded ape/human divide) has been inverted, and humans have fallen from the top branch of species exceptionalism in the diagrams of the evolution of life and are "hunted, caged, and forced to mate by civilized apes" (the film poster's tag line). Arriving in his military spaceship fueled with a sense of discovery, Taylor is the embodiment of what Donna Haraway calls "Man the Hunter": guarantor of a future for liberal democracy armed with technology and an urge to travel as the means to establish an exchange system critical to the free-world ideology of universalist man (1989, 187). Embodying the historicity of the myth of discovery, the temporal clock starts from his arrival. Apes or, in Haraway's terms, the "dark humanoid animal[s]" are figured in the fantasist struggle over future evolution (with a barely coded message about the implications of "evolving" apes for the survival of white humans) (190). Despite the power differential (Taylor has crash-landed, akin to the colonialists that depended on indigenous generosity for their survival), the apes must constantly strive to prove their humanity, even though they are clearly not human but ape. In the slippages, the film forgets its playing in blackface.[8] In the thin analogy to the colonial story, the apes are caught in the dissonance of racial categories of the inhuman in paleontology. The film highlights the sticky oscillation between ideas of absolute difference and cultural graduation in understandings of racial difference in the borderlands of colonial power.

The narrative fold of the half-ape, half-human conflict explicitly engages with the hierarchical categories of the human in subhuman categories, where species equals race and species role reversal attests to a liberal allegory of the fear of racial conflict. While the apes in blackface represent a specific engagement with the racial panic of 1968, amid the civil rights movement and Vietnam War, the racial logics predicated on species sort to engender a politics of tolerance cut with a distinct reasoning against interspecies mixing (read: fear of miscegenation), lest human triumphalism (coded white) fall from its genealogical perch. The apes both adhere to and depart from the position of the ideal paleontological human. Between origin and destination, the fantasy (and implied horror) of apes released from their arrested ancestral development keeps them in their place on the evolutionary ladder to engender the myth of justifiable racial violence. The inversion of the evolutionary horizon (understood through a paleontological/social Darwinist lens) is implicitly used to justify the ongoing violence of racial suppression and segregation, argued through an overwriting of geologic perspectivism. This racial logic posits white universality as a product of geologic determinism told through the achievement of species placement within the geologics of temporal coding of species by way of race. Toward the end of the film, Taylor, coded Western white male, center of the meaning of the universe or, in Wynter's terms, "overrepresented Man" (2003) is arrested before the decomposing Statue of Liberty to encounter geology as a projection and shattering of his psychic needs. In this infamous beach scene, Taylor realizes he is stranded on the terminal beach of time and that the future of humanity has already been destroyed. The future is already the past, and geologic realism presents a dispossession of the subject written on the shore awaiting its effacement, like Foucault's account of man as a historical formation written in the sand. Liberty is washed up as a future fossil. Posed between the West's racial and technological supremacy aspirations laid bare in the half-buried Liberty and a scantily clad mute female on a horse (charged with reproductive futurism), Taylor falls on his knees before the ultimate "mother" of the nation and shouts his desperate lament: "You maniacs, you blew it up!" Geologic realism here offers a recursive teleology where the end is proffered as a new beginning and a warning to white heteropatriarchy about the fall of empire.

"The Terminal Beach" (1964)

In another blown-up island, a fugitive is wandering lost in the ruins and losing himself in J. G. Ballard's short story "The Terminal Beach." Traven, an ex-pilot who has lost his wife and son, is stuck in another kind of the epochal realism as he makes his way to the terminal beach of the story. The terminal beach is the Pacific island of Eniwetok, which has been abandoned by the Atomic Energy Commission for nuclear testing, and where Traven persists, working through the strange logics of this new geologic epoch where the sand and nuclear blasts have formed a newstrata:

> The series of weapons tests had fused the sand in layers, and the pseudogeological strata condensed the brief epochs, micro-seconds in duration, of thermonuclear time. Typically the island inverted the geologist's maxim, "The key to the past lies in the present." Here, the key to the present lay in the future. The island was a fossil of time future, its bunkers and blockhouses illustrated the principle that the fossil record of life was one of armour and the exoskeleton.
>
> *(1964, 137–138)*

The island is told as an affectual architecture of the atomic orb that had blown apart spatial and temporal boundaries, "separating one moment from the next like two quantal events" (140). In parallel to the landscape, Traven is already on the run from time. Like J. M. Coetzee's character Michael K, his investigations, or "research," on the island entail a kind of psychic gardening in the strange geographic ruin of the future (1983, 166). This is what Ballard calls an "ontological garden" against the "world of quantal flux" (153). The structure of megaliths provides the substitute "for those functions of his mind which gave to it its sense of the sustained rational order of space and time. Without them, his awareness of reality shrank to little more than the few square inches of sand beneath his feet" (145).

For Ballard, the atomic age and his proximity to the bombings in Nagasaki and Hiroshima at his internment camp in Shanghai signified the end of a particular form of history and the beginning of a chronological confusion that arranges itself in the age of surfaces and material/temporal recombinations. The atomic is a cut in the consciousness of time and space; the burning atomic light is everywhere at the same time, making nuclear photographs of people on the sidewalks, commandeering the Empire of the Sun, and glazing earthenware pots with boutique Anthropocene radiation signatures. As this atomic signature and its geochemical marker are emblazoned across all Earth, ecologies, and bone, it is the favored origination moment for the start of the Anthropocene, and as such, these fossils are proffered as exhibits in the Anthropocene museum.

There is another archive of this island, of the multiple displacements, relocations, and exiles, recurring generational tumors, miscarriages, and malformed births, body burdens and toxicities that were the forced inhabitation of these bomb events for Marshallese and Pacific Islanders. In the speculative abstraction of these atomic fossils, there is a living archive of personhood subjected to years of military testing, irradiation, and obliteration, whose nonrepresentation is interned in the representational and racial geo-logics of the military "theaters" that constructed global-world-space. Traven is analogous to an ersatz Anthropocene speculative geologist breaking "free from the hazards of space and time" (145) contemplating the arche-fossils of the future, where subjectivity becomes a space of slippage and dissolves in the material cracks of molecular bonds. All the while, another geologic realism lives on the beach through its erasure; it is returned into the official archive in bodies inscripted into the US human testing program that accompanied the atomic "tests." The bomb is the achievement of the advent of geologic time conditions through material transformations in science. This is the terminal beach that fossilizes the future through the present. Its legacy is waste, as fallout, as contamination, as future disposal, and as temporal problematics perennially located in indigenous lands and bodies. This nuclear colonialism represents both an implosion of modernity's technics of time through the ordering of matter and the consolidation of its unconscious fantasies of obliteration, scaled up from other categories (of those outside European ethno class man, as Wynter names it) to whole populations (2003).

The nuclear annihilation that occupies the horizon in Ballard's writing becomes palpable as a subjective mode, where the temporal possibility of unleashing the explosion of nowhere is in constant dialogue with the fractures within subjectivity, unpicking the fragile cohesion of location within time and space. It parallels another implosion in the colonial matrix of time-space in the Middle Passage. Territory becomes crystalline, no longer a destination or accomplishment of conquest but a property of the inhuman dimensions of matter that subjectivity must negotiate in the present tense. Time becomes the territory that is unleashed by the implosion/explosion of material conditions of coloniality. In 1977, the US military began decontamination of Eniwetok and other islands, during which time they mixed more than 85,000 cubic meters of contaminated soil and debris from the islands (including debris from Bikini and Rongelap Atolls) with Portland cement and buried it in an atomic blast crater on the northern end of the atoll's Runit Island (the crater was created by the 1958 "Cactus" nuclear weapons test). The radioactive waste soil and Portland cement is categorized as a classic "Stage 11" Anthropocene fossil, according to the new classifications of mineralogy: "A time when mineral diversity is experiencing a punctuation event owing to the pervasive near-surface effects of human industrial society" (Hazen 2017, 595). As a time assemblage of Jurassic meets Atomic, dead fossil sea matter from the Island of

Portland, on the Jurassic coast in England, was fused with atomic waste from the 1950s tests in the atolls.[9] In 2018, the islands are on the front line of climate change. The Pacific beach is simultaneously a shift in property and properties, in territory and territoriality: property of a neocolonial superpower and dispossessed islanders, and the shifting material properties that organize extinguishment and possibilities across different logics of the present and deep time of the future, as corporeal and incorporeal toxicities.

The lithic/littoral scenes of both *Planet of the Apes* and "The Terminal Beach" play out the cultural arrangements of what geologists have recently called the new mineralogy of the Anthropocene epoch: "index fossils" or the fossilized occurrences around absences made from mining and various forms of material extraction, where the absence of certain concentrations of minerals is an indication of human perturbations. In *Planet of the Apes*, the future fossil is symbolic monumentalism of whiteness. In "The Terminal Beach," the speculative psychic indexing of time and matter after nuclear testing and the material recombinations of its subsequent containment are an index of the militarization of the strata. In the geologic realism of nuclear colonialism, Eniwetok, like so many Pacific atolls used for nuclear "testing" as part of the Pacific Proving Grounds, had its indigenous population expelled by the US Army. In this sense, the island is indexicality linked to the racialized underbelly of the Anthropocene and its origin moments that are defined through the foundational structures of race.[10] There was no such thing as a test; every atomic bomb was an ecocidal reality practiced on personhood and life forms—an ongoing praxis of colonial inhumanism. In another image of geologic realism, the children of Marshallese islanders danced in the falling atomic coral, not recognizing its toxic payload. As the radiation burnt their skin, the US Army waited seventy-two hours before it commenced its evacuation and its human testing program, Project 4.1.[11] Here nuclear explosions and their ongoing effects were not occasions for speculative time travel but were rendered through the intergenerational flesh of indigenous populations. This is Glissant's "painful notion of time," crisscrossing displacement, exile, and segregation.

Daughters of the Dust (1991)

Dispensing with an account of the patriarchal white subject as *the* subject of geologic realism, the final lithology, Julie Dash's *Daughters of the Dust*, claims its own possibilities of radical liberation from racialized subjectivity on the beach of time through a creolized aesthetic. Dash's film explores three generations of Gullah women and the legacy of slavery (marked by the poisonous indigo on the women's fingers) to bring the past into the present to become a future shore, in their Sea Islands home, Saint Helena Island, off the coast of South Carolina. *Daughters* here is discussed not as a resolve or to redeem a better subject identification but as a challenge to the homogenizing and hierarchizing subjective impulses that have long been gathered in the temporal lines of geologic time. Thinking with deep time requires a reworking and rerouting of the linage of genealogy (and its positioning of legitimate kin) as it has been historically rendered through paleontological accounts of time (as a racialized and biocentric mode of subjectivity). For subjects rendered outside of colonial time yet within its subjugating grammars, this engagement with the pluralities of times' material valences has been a source of resistance and liberatory possibility, from the wild mountain ecological knowledges of maroonage to Glissant's "black beach" of volcanic sand that washes dead ancestors up amid the tourists (1990, 121; 205).

"What's past is prologue," proclaims a character at the beginning of *Daughters.* By having her film narrated by "Unborn Child," Dash recasts the monstrous impossibility of the natal occasion under slavery to weave through patterns of persistence and survival in and around the unspoken history of displacement (Hartman 1997, 13). The unborn child recalls Saidiya Hartman's discussion of the figure of the spirit child caught like the returning tide in the breach of slavery, "who dies only to return again and again in a succession of rebirths. The spirit child shuttles back and forth between worlds of the living and the dead because of the stories not passed on, the ancestors not remembered, the things lost, and the debts not paid. The 'come, go back, child' braves the wreckage of history and bears the burdens that others refuse" (2007, 86). The "come, go back, child" belongs to the black archive in the wreckage of inhuman history. In friendship with Brand's extraordinary genealogy of Marie Ursule's diasporic children in *At the Full and Change of the Moon* (1999), Dash's characters populate and fragment the world with their histories up and down the maternal line. *Daughters* portrays a representation of blackness informed by island living, myth, and intergenerational traditions, in contrast to the dispossession of origins in genealogical isolation and commodified generational relations under slavery and the racial calculus of its afterlife (Hartman 2007, 86). The beach is a space of confrontation with exile, maroonage, and the reclamation of a space in the world. Most recently referenced on Beyoncé's album *Lemonade* (2016), Dash's vision presents black life in "the break" between island and mainland, organized around a vibrant representation of black women on screen, fused together by the social realist mode of the LA Rebellion cinema (and the cinematography of Arthur Jaffa). In the first distributed feature-length film made by an African American woman, Dash articulates the legacy of slavery as being together both in brokenness and in the strength of its independent "poethics," to use Silva's term, of radical black feminist tradition (Silva 2014). Moten and Harney name this blackness "in the break," where

> we remain in the hold, in the break, as if entering again and again the broken world, to trace the visionary company and join it. The contrapuntal island, where we are marooned in search of maroonage, where we linger in stateless emergency, in our lysed cell and held dislocation, our blown standpoint, in (the) study of our sea-born variance, sent by its pre-history into arrivance without arrival.
>
> *(2013, 94)*

While *Daughters* evidences the other worlds of maroonage, the unborn child is a challenge to the nonarrival that is scripted into Western genealogical accounts to justify racial subjugation.[12] Wynter in "Black Metamorphosis" saw both maroonage and the ecologies of slave plots as a form of "replanting" culture in place and establishing time outside of plantation time. Dash refutes colonial time by discarding the cinematic tradition of man as the organizing optic and heroic subject, refusing acknowledgment of overrepresented man. Rather than engaging with the production of marginality in relation to that overrepresentation, *Daughters* bypasses those narrative expectations and imaginaries, centering women and intergenerational lives in a turn away from an acknowledgment of what is cast as the representational center in Western filmic traditions. Opening forms of blackness in cinema beyond their opposition, Dash directs languid forms of inhabitation that are resistant and defiant in the context of the overrepresentation of black death. This is an origin story with an unborn narrator that weaves through generations not in a progressive narrative of emancipation but as a grounding to inhabitation through the weft of movement between generations, and specifically, an inhabitation that is not grounded in the monstrous assault on the possibility of black natality that is specific to the forms of dispossession under the conditions of slavery and its contemporary transformation within the carceral state. In the film, kinship shifts, fragmented narratives move through engagements with the landscape, choices are made between the arrival and departure. The island is both cut off and in conversation with the mainland. There is the possibility of images (the character of the photographer), of a genealogy that is not overwritten with black death and white aesthetics. Dash decenters white scopic regimes through narrative forms and temporal occasion, declaring "you are not here."

The brilliance in being in the collective movement of lives, conceived through multigenerational and diasporic relation, is made without the burden of survival as it is offered in its raged half-given, already taken-away version by white supremacy. If Charlton Heston in *Planet of the Apes* represents the masculine pathos of white humanist existence, then *Daughters of the Dust* imagines and repositions black feminist survival in its freedoms beyond the projects of external description (in colonialism and slavery) that have incarcerated blackness in black social death. Dash's descriptive mode involves production of counterimaginaries not just of subjective modes but of relation. As Hartman articulates in her own work,

> Blackness is defined here in terms of social relationality rather than identity; thus blackness incorporates subjects normatively defined as black, the relations among blacks, whites, and others, and the practices that produce racial difference. Blackness marks a social relationship of dominance and abjection and potentially one of redress and emancipation; it is a contested figure at the very center of social struggle.
>
> *(1997, 56–57)*

Remaking the relations of dominance and abjection, the sociality of the beach in *Daughters* is not a space of dispossession but an opening into the world. The flowing white dresses, the trees, the movement of the women's bodies on the beach, all that beauty and freedom is a commitment to the other nonrepresented black archives and forms of subjective life that do not have recourse to, or need of, man's signifying language. The subtle acknowledgment of slave heritages and chattel is the trace of indigo that stains the women's hands as an external historical mark of the intimate proximities to inhuman matter. Dash's black feminist praxis does not just unmask the heteropatriarchy and its affective architectures (as it is not primarily a praxis for the social critique of these formations) but rearranges relation through an inversion of the politics of nonrecognition. Such praxis expands the temporal logics of relation through radically different experiences of materiality as an implicit critique of the regulatory structure of reason in empire, race, and slavery and its violent structures of matter.

The geologic realism of *Daughters* is the quotidian and creative act of survival in an alien land in the afterlife of slavery, survival that is not overly determined in the film's narrative by Middle Passage epistemologies but finds its own narrative place and future in the time-space of what Wynter calls "replantation" (n.d., 502.1). Replantation, Wynter suggests, was the cultural-ecological cultivation of relation to inhuman and nonhuman forms, from planting foodstuffs in slave plots to developing knowledge of plants and vegetation in maroonage, a sense that gave space to imagination and relation through ecology and outside the machine of industrial plantation and its economic botany. It is this geologic realism that is played out by Dash on the littoral/lithic shores of island life, from the mythic African legacy of the bottle tree to the representational genres of the photographic, where a radical textural pattern of time-space reworks the material and metaphoric as inseparable registers, like Glissant's archipelago poetics that occupies the historical legacies of a forced openness to the sea with a chosen opacity, where "green balls and chains have rolled beneath from one island to the next, weaving shared rivers that we shall open up," in which "the boundary, its structural weakness, becomes our advantage. And in the end its seclusion has been conquered" (1990, 206). Dash's aesthetics offer a filmic vision of this poetics of openness, untethered from the horizon of colonial meaning, opening space on the beach for the praxis of reorientation to other temporalities of relation that survival is already attesting to, "in alliance with the imposed land" (7). Enforced inhuman proximity becomes a way and a passage to other kinds of intimacies beyond colonial heteropatriarchy.

Dash speaks to survival in the indices and intramural spaces of colonial power as a material and poetic act, where, to paraphrase Wynter, "senses become theoreticians, crafting an imaginary that replots the coordinates of black life beyond the calculus of the forced horizons and hierarchies of inhumanism" (n.d., unpublished manuscript, 109). In *Daughters*, Dash presents a geologic poetic that stages black feminist fugitivity not as escape but as a geographical extension within place, an opening up of space and time against the enforced immobility and monolingual intent that marks the rubric of black positionality.

On the beach of reason and realism

If the inscription of colonial man on the terminal beach of time has collapsed and been resurrected in the Anthropocene, this is merely a distraction to the structural arrangements that take place within the fallacy of man and his humanism, as the author and ground to Earth revolutions. Man, as it were, is the ideological distraction and legitimating center to the haphazard reorganization of the geochemical fabric of Earth, imagined through the organic/inorganic schism of the inhuman. The inhuman functions as a regime of power through divisions of matter and conciliations of language that bring subjects and matter into the same classificatory schema across metaphysical and geophysical orders. That is, geology becomes a cipher for the work of extraction of subjects and earth. This is the racialized geotrauma of the colonial Anthropocene. If geologic relations are to be examined, a radical interrogation must remain as traumatic as its passage because imaginaries of geologic realism usher in temporal structures that define the spaces and subjective arrangements of and through futurity. As Silva argues in *Towards a Global Idea of Race*, inclusion in global-world-space is the problem, not the solution, because race is inscribed in the production of space from the get-go, and it is only the refusal of those terms that moves toward dismantling the apparatus of raciality and its inhuman politics. Building on Silva's analysis, I have attempted to show how temporality becomes a means to produce mastery over space and matter by looking at the geologic futurisms entwined in the subjective-temporal positions of the Anthropocene. Naming a set of radical environmental changes that are coupled with increasingly extreme social and political formations, these Anthropocenic rifts map into hyperlocal contexts and durations of relation that amplify environmental exposure along a racialized axis. The configuration of the localities of inhabitation matter, particularly in how they recognize the ongoing colonial histories of displacement, exile, and errantry.[13] Thinking with Glissant and his demand for a "nonprojectile imaginary construct" in the formation of global relation, there is a need to both break the map of reason and understand the histories of subjugation that it imposed on subjects and earth through the lens of inhuman matter (1990, 33). As geologists vie to name the moment of Anthropocenic origination, the identity of time and space is being rearranged and reproduced. As Earth is its own ground and invention, ceaselessly churning its own variations on the ontogeny and orogeny of objects, evolving new forms of life and mineral out of the geologic soup, another inhuman story remains in the caesura that needs reanimation. Commenting on Glissant's poetic intent, Mbembe writes:

> The durability of our world, he insisted, must be thought from the underside of our history, from the slave and the cannibal structures of our modernity,

> from all that was put in place at the time of the slave trade and fed on for centuries. The world that emerged from the cannibal structures is built on countless human bones buried under the ocean, bones that little by little transformed themselves into skeletons and endowed themselves with flesh. … The durability of the world depends on our capacity to reanimate beings and things that seem lifeless.
>
> *(2017, 181)*

As geologic realism underpins and contextualizes organic life across all entities, it is also a site of struggle in the designation of choosing which arrangement of life (and nonlife) is deemed to matter. To call for the end of the world is both to call for an end to the white supremacy of the human that has brought such scenes of subjection in its wake and to strike out into the geotraumas of the inhuman (as Earth and race) to bring back from these scenes of extinction a sensibility that might come to define the inhabitation of Earth through a different durational and subjective relation beyond extraction.[14]

Notes

1 The impetus for this work came from an invitation to the Climate Realism Conference at the Media@McGill International Colloquium (organized by Lynn Badia, Marija Cetinić, and Jeff Diamanti). I am grateful for the discussion with the organizers and fellow presenters. I also express my appreciation to Mary Thomas for time travel on the fossil beach and reminding me of Dash's fantastic film and to the wonderful team at *Social Text* for insightful critical feedback, especially Macarena Gómez-Barris.

2 The realism of the "kitchen sink" is an attempt to reposition attention to the overlooked and marginalized dimensions of life in a confrontational realism that speaks to the raw grittiness of life, its everyday class, racial, and sexual politics.

3 "Inhuman" racial substrata are discussed in Yusoff, *Billion Black Anthropocenes*.

4 On the implicit reproduction of white heteropatriarchy in the context of the posthuman, see King, "Humans Involved"; and Broeck, Gender and the Abjection of Blackness.

5 For a more nuanced understanding of the gendering of blackness see Hartman, *Scenes of Subjection* and McKittrick, *Demonic Ground*, 37–64.

6 See Kant, "Of the Different Human Races (1775)" and "Of the Different Human Races (1777)."

7 Wynter and McKittrick, 2015. "Unparalleled Catastrophe"; McKittrick and Woods, Black Geographies and the Politics of Place; McKittrick, "On Plantations, Prisons, and a Black Sense of Place"; Silva, "1 (life) ÅÄ 0 (blackness) = ∞ − ∞ or ∞ / ∞"; Sharpe, *In the Wake*.

8 McHugh, 2000. "Horses in Blackface"; also see Greene, "Planet of the Apes" as American Myth.

9 For a counternarrative, see Sharpe's discussion of resident time in *In the Wake*, 41; and Agard- Jones, "What the Sands Remember."

10 See Yusoff, *Billion Black Anthropocenes*; Davis and Todd, "On the Importance of a Date."

11 DeLoughrey, "Myth of Isolates"; Kuletz, Tainted Desert.

12 For other worlds of maroonage, see Sayers, Desolate Place for a Defiant People; and Thomas, "Marronnons/Let's Maroon."
13 Last, "We Are the World?"; Gómez- Barris, *Extractive Zone*.
14 On the end to the supremacy of the human, see Silva, "1 (life) ÅÄ 0 (blackness) = ∞ − ∞ or ∞ / ∞."

References

Agard-Jones, Vanessa. 2012. "What the Sands Remember." *GLQ* 18, nos. 2–3: 325–346.
Ballard, J. G. 1964. "The Terminal Beach." In *The Terminal Beach*, 134–55. London: Phoenix.
Brand, Dionne. 1999. *At the Full and Change of the Moon*. New York: Grove.
Brassier, Ray. 2007. *Nihil Unbound: Enlightenment and Extinction*. London: Palgrave Macmillan.
Broeck, Sabine. 2012. *Gender and the Abjection of Blackness*. Albany: SUNY Press, 2018.
Clark, Nigel. 2010. *Inhuman Nature*. London: Sage.
Clark, Nigel. 2012. "Everything but the Earth?" *Progress in Human Geography* 36, no. 5: 685–687.
Clark, Nigel, and Kathryn Yusoff. 2017. "Geosocial Formations and the Anthropocene." *Theory, Culture, and Society* 34, nos. 2–3: 3–23.
Coetzee, John Maxwell. 1983. *Life and Times of Michael K*. London: Vintage Books.
Dash, Julie, dir. 1991. *Daughters of the Dust*. New York: Kino International.
Davis, Heather, and Zoe Todd. 2017. "On the Importance of a Date, Or, Decolonizing the Anthropocene." *ACME: An International Journal for Critical Geographies* 16, no. 4: 761–780.
DeLoughrey, Elizabeth. 2013. "The Myth of Isolates: Ecosystem Ecologies in the Nuclear Pacific." *Cultural Geographies* 20, no. 2: 167–184.
Fanon, Frantz. 1952. *Black Skin, White Masks*. London: Pluto Press.
Foucault, Michel. 1966. *The Order of Things: An Archaeology of the Human Sciences*. London: Vintage.
Freud, Sigmund. 1991. "Beyond the Pleasure Principle." vol. 11. In *The Penguin Freud Library*, edited by James Strachey, Angela Richards, and Albert Dickson. Middlesex, UK: Penguin.
Glissant, Édouard. 1990. *Poetics of Relation*. Ann Arbor: University Press of Michigan.
Gómez-Barris, Macarena. 2017. *The Extractive Zone: Social Ecologies and Decolonial Perspectives*. Durham, NC: Duke University Press.
Greene, Eric. 1998. *"Planet of the Apes" as American Myth: Race, Politics, and Popular Culture*. Middletown: CT: Wesleyan University Press.
Haraway, Donna. 1989. *Primate Visions*. New York: Routledge.
Hartman, Saidiya. 2007. *Lose Your Mother: A Journey along the Atlantic Slave Route*. New York: Farrar, Straus and Giroux.
Hartman, Saidiya. 1997. *Scenes of Subjection: Terror, Slavery, and Self-Making in Nineteenth-Century America*. Oxford: Oxford University Press.
Hazen, Robert M., Edward S. Grew, Marcus J. Origlieri, and Robert T. Downs. 2017. "On the Mineralogy of the 'Anthropocene Epoch.'" *American Mineralogist* 102, no. 3: 595–611.
Kant, Immanuel. 2013; "Of the Different Human Races: An Announcement for Lectures in Physical Geography in the Summer Semester 1775." In *Kant and the Concept of Race: Late Eighteenth Century Writings*, translated and edited by Jon M. Mikkelsen, 41–54. Albany: SUNY Press.
Kant, Immanuel. 2013. "Of the Different Human Races (1777)." In *Kant and the Concept of Race: Late Eighteenth Century Writings*, translated and edited by Jon M. Mikkelsen, 55–72. Albany: SUNY Press.

King, Tiffany L. 2017. "Humans Involved: Lurking in the Lines of Posthumanist Flight." *Critical Ethnic Studies* 3, no. 1: 162–85.

King, Tiffany. 2018. "Off Littorality/The Shoals." Abstract of paper presented at the Association of American Geographers Annual Meeting, New Orleans, LA, April 12. aag.secureabstracts.com/AAG%20Annual%20Meeting%202018/abstracts-gallery/15578. Accessed May 10, 2018.

Kuletz, Valerie. 1998. *The Tainted Desert: Environmental Ruin in the American West*. New York: Routledge.

Last, Angela. 2017. "We Are the World? Anthropocene Cultural Production between Geopoetics and Geopolitics." *Theory, Culture, and Society* 34, nos. 2–3: 147–168.

Mbembe, Achille. 2017. *Critique of Black Reason*. Translated by Laurent Dubois. Durham, NC: Duke University Press.

McHugh, Susan. 2000. "Horses in Blackface: Visualizing Race as Species Difference in *Planet of the Apes*." *South Atlantic Review* 65, no. 2: 40–72.

McKittrick, Katherine. 2006. *Demonic Grounds: Black Women and Cartographies of Struggle*. Minneapolis: University of Minnesota Press.

McKittrick, Katherine. 2011. "On Plantations, Prisons, and a Black Sense of Place." *Social and Cultural Geography* 12, no. 8: 947–63.

McKittrick, Katherine ed. 2014. *Sylvia Wynter: On Being Human as Praxis*. Durham, NC: Duke University Press.

McKittrick, Katherine, and Clyde Woods, eds. 2007. *Black Geographies and the Politics of Place*. Toronto: Between the Lines Press.

Meillassoux, Quentin. 2008. *After Finitude*. London: Continuum.

Moten, Fred, and Stephano Harney. 2013. *The Undercommons*. Wivenhoe: Minor Compositions.

Patterson, Orlando. 1982. *Slavery and Social Death: A Comparative Study*. Cambridge, MA: Harvard University Press.

Povinelli, Elizabeth A. 2016. *Geontologies: A Requiem to Late Liberalism*. Durham, NC: Duke University Press.

Sayers, Daniel O. 2014. *A Desolate Place for a Defiant People: The Archaeology of Maroons, Indigenous Americans, and Enslaved Labourers in the Great Dismal Swamp*. Gainesville: University Press of Florida.

Schaffner, Franklin, dir. 1968. *Planet of the Apes*. Los Angeles: 20th Century Fox.

Sharpe, Christina. 2016. *In the Wake: On Blackness and Being*. Durham, NC: Duke University Press.

Silva, Denise Ferreira da. 2007. *Toward a Global Idea of Race*. Minneapolis: University of Minnesota Press.

Silva, Denise Ferreira da. 2014. "Toward a Black Feminist Poethics: The Quest(ion) of Blackness Toward the End of the World." *Black Scholar* 44, no. 2: 81–97.

Silva, Denise Ferreira da. 2017. "1 (life) ÷ 0 (blackness) = ∞− ∞or ∞/. On Matter beyond the Equation of Value." *e-flux*, no. 79. www.e flux.com/journal/79/94686/1-life-0-blackness-or-on-matter-beyond-the-equation-of-value/.

Thomas, Greg. 2016. "Marronnons/Let's Maroon: Sylvia Wynter's 'Black Metamorphosis' as Species of Maroonage." *Small Axe*, no. 49: 62–78.

Wright, Michelle. 2015. *Physics of Blackness: Beyond Middle Passage Epistemologies*. Minneapolis: University of Minnesota Press.

Wynter, Sylvia. n.d. "Black Metamorphosis: New Natives in a New World." Unpublished Manuscript. Schomburg Center for Research in Black Culture, New York Public Library, Institute of the Black World Records, MG 502, box 1.

Wynter, Sylvia. 1984. "The Ceremony Must Be Found: After Humanism." *Boundary 2* 12, no. 3–13, no. 1: 19–70.

Wynter, Sylvia. 2003. "Unsettling the Coloniality of Being/Power/Truth/Freedom: Toward the Human, after Man, Its Overrepresentation—An Argument." *New Centennial Review* 3, no. 3: 257–337.

Wynter, Sylvia, and Katherine McKittrick. 2015. "Unparalleled Catastrophe for Our Species? Or, to Give Humanness a Different Future: Conversation." In *Sylvia Wynter: On Being Human as Praxis*, edited by Katherine McKittrick, 9–89. Durham NC: Duke University Press.

Yusoff, Kathryn. 2018. *A Billion Black Anthropocenes or None*. Minneapolis: University of Minnesota Press.

Yusoff, Kathryn. 2017. "Epochal Aesthetics: Affectual Infrastructures of the Anthropocene." *e-flux—Architecture*. www.e-flux.com/architecture/accumulation/121847/epochal-aesthetics-affectual-infrastructures-of-the-anthropocene/.

Yusoff, Kathryn, and Mary Thomas. 2018. "Anthropocene." In *The Edinburgh Companion to Animal Studies*, edited by Lynn Turner, Undine Sellbach, and Ron Broglio, 52–64. Edinburgh: Edinburgh University Press.

SECTION 3

Realism and the critique of climate, or climate and the critique of realism

7

THE POETICS OF GEOPOWER

Climate change and the politics of representation

Ingrid Diran and Antoine Traisnel

This chapter examines how the rhetoric of climate change has obscured what is *unchanging* in the Anthropocene: the nature of capitalist production and the contemporary regime of its governmentality. If the Anthropocene has become a recognizable "master narrative" in the past few decades, we must ask what narratological dispositif undergirds this new hegemonic discourse.[1] Heeding Amitav Ghosh's (2016) provocation that the realist novel is coeval with the science of probabilities—which for Ghosh renders realism inherently unfit to account for such improbable events as climate change and extinction—we seek to outline the representational strategies by which the Anthropocene's planetary scope secures the place of the "anthropos" at the center of geological change while simultaneously rendering such change largely intractable to human action—and, by extension, rendering it *unrealistic*. Following Ghosh, we understand realism not as a mimetic replication of a real that exists "out there" and prior to its representation, but rather as a probabilistic regime that defines as real what is predictable and thus, in principle at least, *reproducible*. Realism, in this perspective, conflates representation with reproduction. And what does climate change index if not a crisis in reproduction, when Nature, mythically conceived as that which endlessly reproduces (itself), is found realistically unable to reabsorb human action into its reproductive cycles?

This chapter proceeds in three parts. First, we argue that the Anthropocene not just as a planetary crisis but also as a problem of representation—the problem, namely, of representing the planet as a stable and predictable object of knowledge. To theorize the Anthropocene as a problem of knowledge embedded in an issue of power, we trace the lineaments of governmentality at the geological level—or geopower. Second, we suggest that geopower is subtended by a realist mode of figuration whose function is to present Nature as sustainable and self-replicating, and thus entirely open to capitalist exploitation. Third, we show that the dual effect of this realism has been to naturalize and dematerialize its object: Nature. By way of conclusion, we invoke a "natural-historical" modality of dialectical thinking that, far from being extinguished by the now-no-longer-operative distinction between natural and human time-scales, mobilizes this threshold to call into question the realist tenets that undergird geopower.

Geopower

In *The Shock of the Anthropocene* (2016), Christophe Bonneuil and Jean-Baptiste Fressoz lament the inability of preeminent philosophies of science to theorize the event of climate change. They find Bruno Latour especially culpable, charging him with having bought into the myth of a "great divide" between nature and society, even as he denounces this divide as a fallacious byproduct of the unreflective parenthesis that he calls modernity. Instead of "a narrative of blindness followed by an awakening," Bonneuil and Fressoz propose a different narrative, one whose task is to represent "a history of the marginalization of knowledge and alerts, a story of 'modern disinhibition'" (72). To account for this disinhibition—that is, to shift the emphasis from motifs of awakening to mechanisms that have historically lulled collectives to sleep—they call for a "rematerialized and ecologized history of capitalism" (206). Such a history would account for environmental destructions, not as the unwitting aftermath of a modern uncriticality, but as the consequence of a deeply asymmetrical and theoretically mystified exploitation of planetary resources, beginning, indeed, with representing the planet as resource.

What, then, is a "rematerialized and ecologized history of capitalism"? Bonneuil and Fressoz suggest that it involves a *genealogy* of the Anthropocene as Capitalocene, a genealogy that shows that, "after life, it is the Earth as a whole … that simultaneously becomes the object of knowledge ('geo-knowledge') and government ('geopower')" (2016, 80). Let us linger with that last term, "geopower," which echoes Michel Foucault's idea of biopower. Just as biopower marks the explicit introduction of "life" in political and economic strategies, geopower marks the designation of the "Earth" as an object of political intervention. While Bonneuil and Fressoz mention it only in passing, we elaborate "geopower" here as an analytic for thinking governmentality on a geological scale. Understood as such, geopower can be seen to subsume the very relation of organic and inorganic matter into the explicit calculus of power. Thus, where biopower makes legible the politics of man as biological agent (as living being), geopower now brings man as a collective geological force (as species) into question.

Developing this analogy further, we can recall that Foucault developed the concept of biopower in order to account for how regimes of political knowledge and practice (race, sexuality, demography) could be assimilated to forces of nature, and thus governed from within regimes of truth (rather than of right or law). We propose geopower, likewise, as a way of problematizing the *geologization* of political power in modernity. Specifically, we seek to understand how "Earth" has been constituted as an object of power-knowledge, and also how power over it has been reframed in—and finally *as*—the Anthropocene (itself proleptically defined as the "age of man" to be inscribed in the fossil record, in the future past of the present). If, as Foucault suggests, the eighteenth-century emergence of political economy transformed the old "subject of right" into the liberal "subject of interest" or *homo economicus*, we argue that today the discourse of climate change has refashioned this biosubject into a geosubject, or *homo ecologicus*, who has become fully legible as a political actor under neoliberalism.

Who or what is this geosubject? Rather schematically, we identify three ways in which she differs from the subject of biopower, with which she is nevertheless continuous:

- First, the geosubject "appears as an eco-citizen" (2016, 83), subject of and to an environmentality much as the biosubject is, for instance, subject of and to a sexuality. Geosubjectivity therefore recognizes itself in different discourses of truth. For instance, the unconscious manifests not as neurosis but as carbon "footprint"; guilt appears as an economic debt to Earth ("ghost" acres, biomass, carbon credit); desire moves toward ecosystems and futures (in both the temporal and financial sense) and not only bodies and pleasures.
- Second, the geosubject is embedded in a narrative logic of terminality rather than vitality, and its future is measured not as growth but decay. As Bonneuil and Fressoz explain, "the subject of the Anthropocene and of geopower is caught in a 'geodestiny' of 'humanity as a geological force' that is both heroic and unsustainable, arousing both admiration and terror while reinforcing a certain number of socio-environmental injustices under the consensual banner of the species" (83). In turn, we might say that the population is to the biosubject what *species* is to the geosubject—that is to say, what an axis of *saturation* was for the former, a horizon of *extinction* is for the latter: if population asks *how many (more)?*, species asks *how much (longer)?* And whereas sciences of populations count up the number of individuals, sciences of the species countdown to extinction, determining tipping points and other points of no return along the way. Hence the apocalyptic tonality of most anthropocenic discourses.[2]
- Third, and most importantly for our purpose, where "norms" in biopolitics are *naturalized* as functions of the body or Life in general, Nature in the Anthropocene—as the composite of Life and Nonlife—is *normalized*. This normalization, importantly, enables geosubjects to act on (and be acted on by) a planetary object—Nature—whose "changes" are primarily defined as deviations from the norm.

To be clear, we are not suggesting that biopolitics has run its course as an analytic of power in late liberalism; rather, we argue that biopower is supplemented, mirrored, and subtended by geopower. Geopower is comparable, in this way, to what Elizabeth A. Povinelli calls "geontopower," a formation that both presumes and enforces a strict differentiation, not between life and death (*bios* and *thanatos*), but between Life and Nonlife (*bios* and *geos*). Unlike Povinelli's concept, however, geopower does not mark an end to biopolitics (or prove it a flimsy metaphysics)[3] but on the contrary, demonstrates that a politicized distinction between Life and Nonlife does not *precede* the birth of biopolitics but is absolutely coterminous with it.[4] The ontological scission between Life and Nonlife, that is to say, is as integral to *geopolitical* modernity as the ontological consolidation of Life is to biopolitical modernity.[5]

In turn, a critique of biopower necessarily implies and enables a critique of geopower. In such a critique, the concept of "Earth" or even, of "ecology" as something like a metabolism between Life and Nonlife, no longer appears as a given but rather as an epistemic corollary to the governmentality which grasps Nonlife through the grid of normativity. This is true historically insofar as the defining and securing of "raw" materials have been the basis for the global biopolitical economy since the seventeenth century; it is also true conceptually, since a notion of "Life" as that which is to be cultivated, enhanced, or exterminated—in a word, normalized—could only arise by presupposing a Nonlife that stands in *reserve*, to be inventoried, metabolized, and normalized in turn. To the extent, then, that this geo-governmentality is the condition for biopower, we might call it biopower's geological unconscious rather than its ontological other.

What, then, enables this doubling-back of the naturalization of norms into the normalization of nature? We can glimpse an answer in Foucault's seminar on the *Birth of Biopolitics* (2010) when he refers to a principle of "naturality" that presides over the practice of government beginning in the eighteenth century. There, Foucault explains why it is insufficient to say that governmentality in the liberal era operates on the basis of norms that have been naturalized over time, for to stop here would be to presuppose precisely what is at stake in norms: namely, a "natural" quality that is discoverable through the modern arts of governing. Speaking specifically of a transformation at stake in the emergence of political economy, Foucault explains that

> what political economy discovers are not natural rights that exist prior to the exercise of governmentality; what it discovers is a certain *naturality* specific to the practice of government itself [*une* naturalité *propre à la pratique même du gouvernement*]. There is a nature proper to the objects of governmental action. There is a nature specific to this governmental action itself and this is what political economy will study. This notion of nature will thus completely transform [*basculer*] with the appearance of political economy. For political economy, nature is not an original and reserved region on which the exercise of power should not impinge, on pain of being illegitimate. Nature is something that runs under, through, and in the exercise of governmentality. It is, if you like, its indispensable hypodermis. It is the other face of something whose visible face, visible for the governors, is their own action. Their action has an underside, or rather, it has another face, and this other face … is precisely what political economy studies. It is not background, but a perpetual correlative [*corrélatif perpetuel*].
>
> *(2010, 16, translation modified; 18)*

"Nature" transforms here from that which marks an outside to government (as in the pre-political "state of nature") into an unconscious truth and rule (the "nature" of government) of power. Nature is thus the mirror in which governmental practice is reflected as governmental reason. In the category of "nature," that is to say, one finds the hidden logic of normativity as what remains hidden beneath and within politics.[6] Beginning in the liberal period, Foucault suggests, "nature" no longer functions as background but as "perpetual correlative" for the systematization, implementation, and justification of power.

The implication, moreover, is that "naturality" itself (a term Foucault displaces and qualifies apart from its customary noun much as he does "sexuality" from sex and "criminality" from crime) is not itself a state but a normative function; its role is to bring questions regarding the outer limits of governmental practice inside the scope of that practice itself (or as Foucault puts it elsewhere, to turn the question of jurisdiction into one of *veridiction*).[7] As natural, relations of force become normalized, reproducible and predictable. This means that they are not only are taken for granted, but more importantly are able to generate and demarcate a viable or realistic field of possible intervention. In other words, the naturalization of norms transforms them into the setting for everyday life but, we add: this naturalization takes place entirely at the expense of an exteriority that is not always already continuous with a normative field. Paradoxically, then, "naturality" functions as an apparatus of dematerialization.[8]

If normativity injects a theory of the natural that makes it "hypodermic" to power, it is also true that naturality, from beneath the skin of power, delineates a horizon of intervention in which governmental practices of inventory, speculation, and consumption (or preservation) begin to circulate as truths. Anna Tsing (2015) has termed this liberal epistemology of matter "scalability," or the "ability of a project to change scales smoothly without any change in project frames." Its iconic site, for Tsing, is the European colonial plantation (not coincidentally, a material condition for the science of political economy that Foucault discusses above). "Scalability," she writes, "requires that project elements be oblivious to the indeterminacies of encounter; that's how they allow smooth expansion. Thus, too, scalability banishes meaningful diversity, that is, diversity that might change things … the nonscalable becomes an impediment" (53). Scalability is primarily a method, though it morphs into an ontology: "Portuguese planters stumbled on a formula for smooth expansion. They created self-contained, interchangeable project elements, as follows: exterminate local people and plants; prepare now-empty, unclaimed land; and bring in exotic and isolated labor and crops for production" (39). Through a combination of "alienation, interchangeability, and expansion," these planters nevertheless "shaped the dreams we have come to call progress and modernity," allowing it to be "posited that everything on earth—and beyond—might be scalable, and thus exchangeable at market values" (40).

In this conflation of "naturality" and capitalist abstraction, we glimpse not only the fundamental continuity between liberalism's *homo oeconomicus* and the neoliberal *homo ecologicus*, but also the relationship between the two as a scaling up from the "naturality" that places governmental practice and self-interest beyond rebuke, to the new "naturality" of landscapes depleted by that same self-interest.[9] In this new presumption of devastation—of finding blasted landscapes and ruins of scalability—the ecosubject is simultaneously consigned to and barred from a natural world. To the extent that the "nature" of this world is already ruined, or even unnatural, the geosubject can no longer distinguish between nature and normativity, since it is now not only impossible to refer to norms without the correlative of naturality, but it is also impossible to refer to naturality without the correlative of normativity or its absence (that is, as reference to what *was* natural but is no longer). Climate change emerges as a problem when normativity, scaled up to the meteorological and geological levels, diagnoses a deviation perceptible in a material and nonhuman world that no longer reproduces the norms that the notion of "naturality" created. It is not Nature's capacity for self-reproduction that climate change disrupts, in other words, but the reproduction of normativity that was thought to inhere therein.

The climate of fiction

All this, we suggest, has had a distinct effect on Anthropocenic discourse, for Nature now circulates under forms of representation that have entirely obscured their status as normative implements—that is, as interchangeable vectors of self-replication and expansion—and tend instead to pass for projected or probable truth, for forms of realism.

It is useful here to remember that Foucault turned to genealogy in order to short-circuit this loop of realist representation and the reproduction of reality in its image. Indeed, in his genealogy of criminality, he aligns the transition from the epic to the novel with the moment "when the normal took over from the ancestral, and measurement from status, thus substituting for the individuality of the memorable man that of the calculable man" (*Discipline and Punish*, 193). Elsewhere, he contrasts his genealogical method more broadly against the concept of genesis: "as opposed to a genesis oriented towards the unity of multiple descendants, what is proposed instead is a *genealogy*, that is, something that attempts to restore the conditions for the appearance of a singularity born out of multiple determining elements of which it is not the product, but rather the effect" (2007a, 64). In other words, genealogy is offered as a corrective to the smooth plane of origin stories; defining conditions for the apparition of a "genesis," genealogy identifies the "singularity" and contingency of the plane on which such stories might unfold. Genealogy treats the order of discourse itself on the order of an event.

If, as D.A. Miller (1988) suggests, the realist novel is a testing ground for representing social normativity and for normalizing social representations, we should note that in the narrative universe of these novels, "genesis" and "descendants," retain pride of place—a fact that requires the background, atmosphere, and indeed the *climate* of the story to approximate a self-replicating plane of appearance. To take a concrete example, let us return to Ghosh's (2016) suggestion that the realist novel is the paradigmatic site for studying the naturalization of norms (and the attendant normalization of nature). Ghosh's observation that the "rhetoric of the everyday" we associate with realism appears at the exact time "a regime of statistics … was beginning to give new shapes to society" (19) is anticipated in D. A. Miller's contention that the novel is a regulatory apparatus and police technology whose vocation is "to enforce not so much a norm as the normality of normativeness itself" (Miller 1988, 18).[10] This is achieved through what Miller calls the "coherence of delinquency," a regular set of features which establishes crime not as anomalous or episodic so much as *abnormal*. Crime, that is to say, retains its own systematicity. It is governed by an abnormality that imparts a routine visibility that gives a face or profile to the criminal and the recognition of which sanctions a certain paranoia. As Miller suggests, in realist texts, what had hitherto "seemed natural and commonplace" (29) becomes suspicious precisely insofar as abnormalities become visible and rational. The function of the detective in particular is to notice and recognize the norm within the abnormal: if "normality has been hitherto defined as a matter of *not needing* the police … [t]he investigation repairs this normality, not only by solving the crime, but also, far more important, by withdrawing from what had been, for an aberrant moment, its 'scene'" (3).

What is scandalous about crimes in realist novels, then, is not their transgressions of the norm, but their normalization of transgression itself. To this extent, such narratives no longer require police intervention, but only the deductive powers of the amateur detective. The detective can "think like a criminal," outlining the rationality of deviance. As Miller writes, "Unobtrusively supplying the place of the police in places where the police cannot be, the mechanisms of discipline seem to entail a relative *relaxation* of policing power. No doubt this manner of passing off the regulation of everyday life is the best manner of passing it on" (1988, 16, italics in the original). What Miller calls the "genetic organization" of the realist novel is therefore premised upon a naturalization of norms (beginning with that of the abnormal): "Structured as a genesis," Miller writes, "the narrative that seems to resist a novel's control … becomes a technique for achieving it" (27).

Ghosh, who writes from the perspective of a novelist who has himself experienced the probabilistic limits of the form, echoes Miller's argument by invoking its converse. He explains, that is, what it is like to not "achieve" a novel on account of events that fail to adhere to the "coherence of delinquency." Having lived, "on the afternoon of March 17, 1978 [through] the first tornado to hit Delhi in recorded meteorological history," Ghosh reports wishing, on numerous occasions, to "put [this experience] to use in a novel, only to meet with failure at every attempt." He ponders:

> On the face of it there is no reason why such an event should be difficult to translate into fiction; after all, many novels are filled with strange happenings. Why then did I fail, despite my best efforts, to send a character down a road that is imminently to be struck by a tornado? In reflecting on this, I find myself asking, what would I make of such a scene were I to come across it in

> a novel written by someone else? I suspect that my response would be one of incredulity; I would be inclined to think that the scene was a contrivance of last resort. Surely only a writer whose imaginative resources were utterly depleted would fall back on a situation of such extreme improbability?
>
> *(2016, 16)*

The *improbability* of the Delhi tornado amounts to its novelistic *impossibility*. A literary "contrivance," the tornado resists narrativization not because it could not or did not take place, but purely because it *should not*. As an almost unrepeatable event, the tornado's exceptionality is neither routine nor regular enough to be assimilated to abnormality. It cannot pass for a regular windstorm "scaled up." For this reason, the altogether factual tornado threatens a novelistic regime that adheres not to what actually happens, but to what is reproducible—and hence, normalizable—in abnormality itself.

Ghosh speculates that his tornado might have had better literary luck had it taken place a few centuries earlier, before dictates of realism held sway, when extraordinary happenings formed the very principle of storytelling: "Before the birth of the modern novel, wherever stories were told, fiction delighted in the unheard-of and the unlikely. Narratives such as those of *The Arabian Nights*, *Journey to the West* and *The Decameron* proceed by leaping blithely from one exceptional event to another. Novels too proceed in this fashion, but what is distinctive about the form is precisely the concealment of those exceptional moments that serve as the motor of narrative" (2016, 16). Exceptional events, importantly, are for Ghosh a point of continuity between premodern literature and modern novels; whereas the former delights and finds in anomalies a *raison d'être*, the latter guards them like a shameful family secret, like the very specter of illegitimacy.

Ghosh turns to Franco Moretti's concept of "filler" to explain how this "concealment" functions aesthetically in modern literature: "fillers are an attempt at rationalising the novelistic universe: turning it into a world of few surprises, fewer adventures, and no miracles at all" (quoted in Ghosh 2016, 19)—in a way that recapitulates and expands upon Miller's point about the "coherence of delinquency" (Miller 1988, 4). Whereas Miller investigates such coherence as an imaginative elaboration of social discipline, Ghosh notes the force that normalization exerts as a policing of the imagination itself that novelists impose upon their own text. If Foucault once mused that modern social discipline achieved the unlikely feat of rendering prisoners their own warden and criminals their own judge, we might say that, no less improbably, the modern novel has rendered writers their own critic. Guilty from the first, the novelist attempts to expiate the reality of aberrant aberrations from fiction, parading out a rational novelistic universe at the expense, often, of reality itself. Therein lies what Ghosh calls "the great derangement."

And yet if exceptional events remain "the motor of narrative" even in the literature defined by their concealment or disavowal, in what way do anomalies like Ghosh's tornado, and other eventive "non-scalables," feature within these texts? If the modern novel still relies on singularities, how and where are they? Moretti's term "filler" raises the question of what guarantees narrative *emptiness* in the first place. If one answer is atmosphere and tone, those ambient effects and totality of particulars that comprise the regularity of narrative, then we might say that the realist novel owes its story to a *stable climate*. But this equally implies that such stories cannot but face an insurmountable difficulty in depicting a phenomenon like climate *change*, where norms are not merely deviated from but themselves shown to be "deranged." Climate change, like Ghosh's tornado, is antithetical not merely to a believable plot, but to the "genetic organization" that secures the conditions for believability in the first place. Disrupted is the stable climate, the plane of appearance, the "naturality," in which narratives unfold and meaning develops.

What these observations begin to outline is a structural limit of realist narrativity, one brought into focus and into crisis by the reality of climate change. This limit concerns not what "happens" in a story but rather the normalized and empty horizon—the implied naturality—within which events take place. Though stories of the unbelievable or miraculous have long been told in minor genres such as "the gothic, the romance or the melodrama … fantasy, horror, and science fiction" (Ghosh 2016, 24), these stories have also tended, continues Ghosh, to "court eviction from the mansion in which serious fiction has long been in residence." They remain stories marked by a certain *irrealism*.

In what exactly does this irrealism consist? For one, it appears to be a characteristic of their nonscalability, which is not the same as ineffability or narrative impossibility. Tsing's book, *The Mushroom at the End of the World: On the Possibility of Life in Capitalist Ruins* (2015), although not itself fictional, is a case in point: both an object of her analysis (matsutake mushrooms) and the mode of that analysis (empirical research) are explicitly defined as nonscalable, responding to and growing out of the various failures of scalability—the mushroom in and on the ruins of the timber plantation, the anthropologist in and on the remains of abstraction, a theory among the remains of generalizations. Tsing declares (in a sort of echo or anticipation of Bonneuil and Fressoz) that "the time has come for new ways of telling true stories beyond civilizational first principles," defined later as "alienation, interchangeability, and expansion" —or in a word, scalability. And "without Man and Nature," she continues, "all creatures can come back to life … no longer relegated to whispers in the night, such stories might be simultaneously true and fabulous. How else can we account for the fact that anything is alive in the mess we have made?" (viii). Describing her book as "a riot of short chapters" and an "open ended-assemblage … mimicking the patchiness of the world I am trying to describe," Tsing locates her mimetic impulse not in content but form: that is, in a practice of attention. Later she advocates the scientificity of her choice: "to listen to and tell a rush of stories is a *method*. And why not make the strong claim and call it a science, an addition to knowledge? Its research object is contaminated diversity; its unit of analysis is the indeterminate encounter" (37; italics in the original). With an always contaminated specimen for a research object, and a unit of analysis that is useless as one of scalable measure, patchiness characterizes both the impossibility of *cultivating* matsutake—"matsutake resist the conditions of the plantation";

"matsutake foragers … are independent, finding their way without formal employment" (40)—and the nonviability of that narrative monoculture known as realism.

Thus, it is no exaggeration to say that the matsutake, like the rush of stories told about it, becomes both more and less than itself in the course of its history. "The fruiting bodies of an underground fungus" (Tsing 2015, 40) not only sustain their host trees but also enable the subterranean flowering of theoretical insight without the relation between these two blooms being interchangeable, homologous, or normalized. Their relation is a salvaged gift, born of entanglement and patchiness—*pastiche*—rather than expansion. For Tsing, the translation of mushroom into a monograph involves foraging, just as foraging requires a gathering of observable bodies come to fruition. Perhaps there is, in other words, not just a science to be gleaned from collaborative survival and entanglement, but a poetics and imaginary too.

The margin of irrealism in non-serious literary genres, we are suggesting, might be rethought with Tsing's example in mind as an out- or overgrowth of something else, a counterfeit or catachrestic effect of torsion within narrative that turns it back upon itself, tropes itself, renders itself parabolic or, in Tsing's terms, "simultaneously true and fabulous" (2015, viii). This internal displacement of narrative, a hollowing out of truth that is also a welcoming in of the fable and the allegory, might (in contrast to the regulatory poetics of realist norms) be described as a kind of narrative shaping. Such shaping characterizes a tradition of literature that unapologetically foregoes Ghosh's "manor" of literary seriousness in favor of mannerism, play, and change. To shape narrative in this way is not simply to defy belief or the believable; rather, it is to act both in and on conditions of believability which are already exhausted. In other words, it is to write in and on a narrative impossibility. It is to begin with the tornado.[11]

Doing so nevertheless requires suspending a search for the "coherence of delinquency" in a way that embraces the singular once again as a "motor of narrative," albeit one that stalls and stutters at least as much as it runs. Shaping, as narrative practice, works on what it finds, manipulating the narrative plane of appearance and not merely figures that appear upon it. For this reason, the embrace of change in and on narrative—the sort of manipulation that non-serious genres like science fiction and twice-told tales have always enacted—cannot but comprise paradigms for a poetics of climate change.[12]

Representation/naturalization/dematerialization

We have been arguing that the Anthropocene is a master term for the realist regime of geopower whose normative force operates under the alias of the Anthropos, and which not only obscures its own contingency, but also ensures its reproducibility in the guise of a dual realism (as statistical *likelihood* on the one hand, as *verisimilitude* on the other). We have sought, in turn, to ask: what are the "multiple determining elements" (Foucault 2007b, 64) out of which this singularity called Anthropos has been produced as a truth-effect? And what method is suited to "restoring conditions for the appearance" of something that has so swiftly passed for the very essence of the human and the basis for governing the planet?

In biopolitics, we recall, the normalized element and prevalent truth-effect is Life. Framed as a transindividual and transpecial current, Life enables a variety of acts and inactions to be reduced to labor power or the potential to produce. Paolo Virno's idea that, in late (post-Fordist) liberalism, capital is "the commerce of potential as potential" (2004, 84) is instructive here because Life as labor power means that there is no longer living without working, and virtually every instance of the former appears as an expression of the latter. Thus, to the extent that Life retains its "naturality," so too does work. With the advent of geopower, the normalized element and prevalent truth-effect is "Nature" or "Earth," defined in this case as a normal and reproducible relation between the living and nonliving. As compared to the biopolitical schema, productive potentiality here is framed *in the passive*: Nature is an ability to be worked (or equally to be set aside, conserved, or offset). Integrated into capitalism's calculus of potentiality, the Earth is thought to be entirely commensurable with the value form. In this sense, the "commerce of potential as potential" precisely *commercializes* the interchange between Life and Nonlife (and converts environmental crises into "innovative" business opportunities).

Furthermore, the normalizing techniques of geopower enable it to constitute the very field over which it governs and posit an Anthropos as a principal and originary agent in modernity's (realist) story. The Anthropos as origin story for the present, in other words, is hardly indicative of the "negative universal history" that Dipesh Chakrabarty advocates we adopt from his influential 2009 essay "The Climate of History: Four Theses." On the contrary, this species-wide story of the Anthropos—however negatively defined or evacuated of positive contents—is a story of power scaled up to the planetary level. While Chakrabarty claims that Marxist dialectics no longer work as well when the distinction between geological and human time has collapsed (2009, 199), we would suggest that there is in fact an abiding importance to dialectical materialist theory as a way of reversing these mystifying vectors of power.[13]

The reversal we hope to leverage with regard to geopower is crystallized (if not fully elaborated) in terms like Capitalocene, wherein inorganic nature—stone and sky, *geos* and *meteoros*—are not taken as natural givens, but as indices of history. More precisely, we seek to view how these categories have operated within normative frameworks. In this way, our proposal is consistent with the theoretical gesture suggested by Bonneuil and Fressoz's idea of ecohistory as part of a materialist critique of the Anthropocene and its fables of awakening. If a dematerialized Anthropocene lends overproduction and geo-debt outside the Western world an air of inevitability, rematerialization draws attention to concrete mechanisms—including, we would suggest, epistemic techniques of normativity—that remain open and contestable.

But perhaps there is even more at stake in the dialectical project of rematerialization: if the Anthropos assumes the place of liberal universal subject—and may in fact be the same thing—what also flashes before us is the way in which geopower has always been coextensive with, and implicated within, biopower. We remember, for instance, that species was a determining category in Foucault's conceptualization of the biosubject (the subject of life), and that in *The Order of Things* he traces the invention of Life to Cuvier's work in comparative anatomy—a work fully contingent on the new sciences of geology and paleontology (themselves propelled by the colonial excavation of minerals and the capitalist excavation of fossil fuels).[14] To the very extent, then, that the Anthropocene can be said to begin with the carbonization of the economy as what Andreas Malm has termed fossil capital, the category of the Anthropos has been born from proleptic contemplations of humanity's remains, an imaginary of *capital fossils*. In other words, biopower's vitalist tendency shows itself to be a negative image—an allegory—of the sciences of transience and decay: paleontology and geology.

Pace Chakrabarty, then, the human *can* experience itself as species, but only in the future perfect, as what it *will have been*, in the shape and rhythm of its extinction. Thus, we should heed Deleuze when he invites us to take literally—or perhaps, *lithically*—Foucault's prophecy at the end of *The Order of Things* as he presents Man as a face of sand about to be effaced by the tide.[15] This, precisely, is the future-anteriority of the Anthropocene, a projection of remnants. In this regard, what Walter Benjamin says of the mourning play—that "history merges into the setting" and thereby composes a "natural history," where nature is no longer antithetical to history—might apply equally to this scene of the Anthropocene, in which "chronological movement is grasped and analyzed in a spatial image" (1998, 92). We are now faced with asserting what a Capitalocene cannot permit at any cost: the materiality and spatial image of nature not as self-reproducing or self-resembling norm but as a self-differing rhythm of transience. This is where geopower indexes not a realism, a master narrative, but an allegorical Anthroposcene of mourning.

The work of Bonneuil and Fressoz—like those of Jason Moore and Andreas Malm—renders manifest the way in which the symbolism of nature has been construed by capitalist logic as a self-sustaining reiteration of the same. We have sought to situate the emergence of this logic within the history of normativity and its attendant narratives and realist imaginaries. Doing so allows us to see in the changing face of nature around us, the unchanging face of exploitation and extraction: the *petrification* of capital as the real subsumption of Life and Nonlife. This symbolic construal of nature also predisposes it to the realism of probability and verisimilitude—in a word, to the realism of a norm. But when, as in the current climate crisis, the material bedrock of nature is in myriad ways upturned, it cedes its realism and normativity, exhibiting itself as changeable face of power.

Notes

1 In *La Part inconstructible de la terre*, Frédéric Neyrat emphasizes the representational issues raised by the Anthropocene as a master narrative of a new kind, both "myth" (founding

its legitimacy in an originary sin) and "metanarrative" (justified by the promise of a future remedy). The Anthropocene, Neyrat argues, is an empty signifier over whose signified an ideological battle is waged by the scientists, politicians, and businessmen he calls the "geo-constructivits." Instead of focusing on which account is the most adequate or convincing, Neyrat invites us to question the mechanisms by which we might be tempted to accept the premise of this narrative that consecrates the anthropos as the new "master of climate" (2016, 67–86).

2 While we cannot treat this problem here, the stakes of Anthropocenic eschatology merit a fuller elaboration.

3 Povinelli (through a reading of Badiou) maintains, that Foucault has been exposed as a metaphysician who himself naturalizes geontopower in order to conceptualize biopolitics (2015, 7–8).

4 Povinelli introduces "geontopower" to make intelligible the distinction between Life and Nonlife that biopolitics, she argues, takes as an ontological given. While we agree with Povinelli's analysis of the critical role played by geontopower in the Western imperial project and especially in "late settler liberal" governance, we depart from her idea that geontopower constitutes a metaphysical aporia in Foucault's analysis of biopower and thus comes as an absolute shock to that theoretical system.

5 One might indeed argue that, scientifically, there is no such thing as Life, conceived as a unified plane of commensurability, before the invention of biology by Cuvier. There were living beings, to be sure, but no Life as their common denominator. On this, see Foucault (2007a, 139), and Lynne Huffer (2017).

6 Phillipe Sabot has discussed the manner in which technologies for determining "norms" increasingly resemble a Spinozan immanent cause in the later liberal period, and for this reason, can be described less as an "action on," than an "action with" the norm's own field of intervention (2016).

7 See lecture of January 10, 1979 in Foucault (2010).

8 The effect of capitalist geopower on the planet—as upon the animal and the human laborer—is not merely estrangement, but *dematerialization* of both activity and inactivity. We think here of theorists like Cesare Casarino, Hardt and Negri, and others, who show that "biopolitical production" has a propensity to present itself as "immaterial" when it is, in effect, anything but. Just as the "vital" functions of the worker are, under biopolitics, subsumed under his potential value as labor, so too are the "latencies" of the earth commodified, including in their conservation, waste, or disuse. And just as, under biopolitics, productivity appears not just natural but indispensable to our very heartbeat (e.g. the "fitbit" watch) so too is a certain potential to produce (including under the heading of the "renewable" resource) nowadays indistinguishable from any concept of the Earth whatsoever.

9 *Homo ecologicus* is also continuous with the neoliberal variant of *homo oeconomicus* as Foucault describes him, namely, as an investor and entrepreneur of the self. However, as we indicate above, "ecological" investments are framed primarily as debts, not credits. In a companion piece to this essay titled "The Birth of Geopower" (*forthcoming* with *Diacritics*), we trace the mutations of geopower under neoliberalism. Under liberalism, we argue, the norms of Nature were able to function as a *given*, such that the space of naturality functioned as a vital reserve for both the life of the nation and the accumulation of capital. The energy crisis of the 1970s, however, announced the finitude and limits of this reserve, both as a material reality and a mode of thought. Neoliberalism responds to this non-reserve by reconceiving finitude itself as a *project* and investment. In other words, the very degradations of the present take measure of the promise the future: the worse things get, the better they can become. This promissory project, which informs the speculative

logic of neoliberalism, has evolved into a logic of governance, which, in addition to privatizing risk and responsibility, profits off of these debts to the future.

10 We do not have the space to develop this here, but it should be noted that this observation does not just apply to the novel but, conceivably, to all artworks that have integrated reproducibility as their operating principle. In "The Work of Art" essay, Benjamin already noted the new significance of statistics in the theoretical sphere: "The alignment of reality with the masses and of the masses with reality is a process of immeasurable importance for both thinking and perception" (1998, 256).

11 Remember that Benjamin construes the origin no longer in genetic terms as a coming into being (origin as *Entstehung*) but as "a *whirlpool* in the river of becoming [that] swallows up the genetic material in its rhythm" (origin as *Ursprung*). The contrast between symbolism's continuist aesthetics and allegory's catastrophic and fragmentary poetics articulated in the *Trauerspiel* book provides an inside into the concept of "natural history" we borrow from Benjamin for our conclusion.

12 In a sequel to this essay, we will examine in greater detail the role played by shaping in two irrealist narratives, Nathaniel Hawthorne's "The Great Stone Face" and Octavia Butler's *The Parable of the Sower*, both of which we read as emblematic of a minor literature that rejects the "genetic organization" of realist novels and foregrounds the changing climate as its protagonist.

13 For a scathing critique of Chakrabarty's genetic narrative, see Malm and Hornborg (2014).

14 As an unintended consequence, Povinelli reminds us, "the exploitation of the coalfields also uncovered large stratified fossil beds that helped spur the foundation of modern geologic chronology" (2015, 10). See also Kathryn Yusoff's *A Billion Black Anthropocenes or None* on the role of colonial-capitalist mining in the emergence of what she terms White Geology.

15 "We must take quite literally the idea that man is a face drawn in the sand between two tides," writes Deleuze: "he is a composition appearing only between two others, a classical past that never knew him, and a future that will no longer know him." It should be noted that, whereas Deleuze discerns in Foucault's image less the tragic disappearance of living men than the advent, spelled by the development of third-generation computers and information science, of a new figure composed by the union of man with "silicon instead of carbon" (1986, 89), the calamitous tonality of most anthropocenic discourses marks instead the profound anxiety in the face of the now *probable* disappearance of the human species due to its dependency on hydrocarbons. In this regard, Deleuze doesn't get it wrong so much as momentarily overlook the persistent indispensability of carbon for the advent of a silicon anthropology.

References

Benjamin, Walter. (1928) 1998. *The Origin of German Tragic Drama*. Translated by John Osborne. New York: Verso.

———.[CI: this is Benjamin again as above] 1968. "The Work of Art in the Age of its Technological Reproducibility," in *Illuminations*, edited by Hannah Arendt. Translated by Harry Zohn. New York: Schocken Books.

Bonneuil, Christophe and Jean-Baptiste Fressoz. 2016. *The Shock of the Anthropocene. The Earth, History, and US*. Translated by David Fernbach. New York: Verso.

Chakrabarty, Dipesh. 2009. "The Climate of History: Four Theses." *Critical Inquiry* 35(2): 197–222.

Deleuze, Gilles. 1986. *Foucault*. Translated by Sean Hand. Minneapolis: University of Minnesota Press.

Foucault, Michel. (1970) 2007a. *The Order of Things. An Archaeology of the Human Sciences*. New York: Routledge.

———. 2007b. "What is Critique?", in *The Politics of Truth*. Translated by Lisa Hochroth and Catherine Porter, edited by Slyvère Lotringer. New York: Semiotext(e).

———. 2010. *The Birth of Biopolitics: Lectures at the Collège de France, 1978–1979*. Translated by Graham Burchell, edited by Michel Senellart. New York: Palgrave Macmillan.

Ghosh, Amitav. 2016. *The Great Derangement: Climate Change and the Unthinkable*. Chicago: Chicago University Press.

Huffer, Lynne. 2017. "Foucault's Fossils: Life Itself and the Return to Nature in Feminist Philosophy." In *Anthropocene Feminism*, edited by Richard Grusin, 65–88. Minneapolis, MN: University of Minnesota Press.

Malm, Andreas & Alf Hornborg. 2014. "The geology of mankind? A critique of the Anthropocene narrative." *The Anthropocene Review* 1 (1): 62–69.

Miller, D.A. 1988. *The Novel and the Police*. Berkeley, CA: The University of California Press.

Neyrat, Frédéric. 2016. *La Part inconstructible de la terre*. Paris: Seuil.

Povinelli, Elizabeth. 2015. *Geontologies: A Requiem to Late Liberalism*. Durham, MA: Duke University Press.

Sabot, Phillipe. 2016. "De Foucault à Macherey, penser les normes." *Methodos* 16: 1–11.

Tsing, Anna Lowenhaupt. 2015. *The Mushroom at the End of the World: On thePossibility of Life in Capitalist Ruins*. Princeton, NJ: Princeton University Press.

Virno, Paolo. 2004. *A Grammar of the Multitude: For an Analysis of Contemporary Forms of Life*, translated by Isabella Bertoletti, James Cascaito, and Andrea Casson. New York: Semiotext(e).

8

PERPLEXING REALITIES

Practicing relativism in the Anthropocene*

Barbara Herrnstein Smith

This chapter is prompted by a question posed by the conveners of the 2017 Climate Realism colloquium and by related views made explicit at the event. "How is realism," they asked, "—in both the aesthetic history of representation and the philosophical tradition that underwrites it—transformed by contending with our new experience of climate in the Anthropocene?"[1]

My focus is on that philosophical tradition: specifically, on the limits of classic realist and related representationalist understandings of knowledge and human cognition, and on alternative constructivist-pragmatist accounts, often labeled "relativist," that, I suggest, offer more useful perspectives on some of the conceptual and practical perplexities presented by climate change.[2] Consideration of three such perplexities will illustrate the point. One is how to respond to malign appropriations of critiques of standard accounts of scientific knowledge developed in fields such as the history and sociology of science. An essay by Bruno Latour often cited in that connection, "Why Has Critique Run Out of Steam?" (2004), provides a point of entry here. A second is how to understand climate "denialism," that is, the rejection of the descriptions, explanations, and predictions of climate science by large segments of the public. Here Ludwik Fleck's classic account of belief systems, *Genesis and Development of a Scientific Fact* (1979 [1935]), is illuminating, especially when joined with contemporary constructivist-pragmatist understandings of human cognition. A third perplexity for dealing with climate change is the existence of multiple, sometimes radically divergent, operative realities. Here the commitment to methodological symmetry associated with constructivist social science can be instructive for ethical and political practice. At

* "Perplexing Realities: Practicing Relativism in the Anthropocene" was originally published in Barbara Herrnstein Smith, *Practicing Relativism in the Anthropocene: On Science, Belief, and the Humanities*, Open Humanities Press, London, 2018. It has been edited for publication in this volume.

the end of this chapter, I reflect on a set of related problems, both conceptual and practical, in recent writings by environmentalists.

I

Realism, in academic philosophy, is the view that the objects we perceive and describe, with the features we observe and characterize, exist independent of our perceptions and ways of describing them. This view is commonly attended by two other ideas. One is that a belief or statement is true insofar as it corresponds to that objective reality or represents its features accurately. The other is that science is a distinctive method for discovering those features and for arriving at true beliefs about their workings. Thus philosopher of science Philip Kitcher, explaining the "broadly realist conception of science" he seeks to defend, describes its components (or, as he writes, its "old-fashioned virtues") as follows: "Scientists find out things about a world that is independent of human cognition; they advance true statements, use concepts that conform to natural divisions, [and] develop schemata that capture objective dependencies" (Kitcher 1995, 127).[3]

These views, formulated in more or less these terms, are familiar. They have, however, been subject to significant challenges. Skeptical philosophers since antiquity have pointed out that we cannot glimpse a putatively autonomous reality around the corner of our own perceptions.[4] Of course, we (or the scientists among us) may chart the apparent regularities of our humanly shared perceptible world and, with various technologies, detect remote, complex, ordinarily imperceptible phenomena. And we (or the philosophers among us) may attempt to infer the essential, underlying features of such a putatively objective reality through, as it is said, the use of reason and logic. But what we (scientists and philosophers included) cannot do is conceive or describe the putative features of such a reality independent of any humanly derived concepts or discursive idioms. Ontologies—purported accounts of what's really out there, or what's underneath it all, or what there really *is*—can be more or less conceptually congenial and pragmatically workable, and we may admire, endorse, or adopt various of them accordingly. What they cannot be, however, is "objectively correct."

Contemporary philosophers have mounted important related challenges to conventional realist views. The work of philosopher of science, P. Kyle Stanford, is of particular interest here. Arguing from the historical record, Stanford advocates an essentially pragmatist (he calls it "instrumentalist") view of the sciences and, with it, of our relation to the experienced world (Stanford 2006, 2016). Scientific theories are effective, he writes, not, as realists commonly maintain, "because they are true" but, rather, "because they help us successfully navigate the world in productive and systematic ways" (2016, 96).[5]

In response to such challenges, academic philosophers have defended, qualified, and updated what they continue to name "realism." Over the past few decades, for example, we have been offered "scientific realism," "realism with a human face," "post-positivist realism," and "speculative realism," each with a set of related

Un missionnaire du moyen âge raconte qu'il avait trouvé le point
où le ciel et la Terre se touchent...

FIGURE 8.1 A glimpse of reality?

ontologies and epistemologies: that is, conjectures about what really *is* and claims about how we (or some of us) can know it. Specific realist and neorealist positions are not my concern here, but I would make a general observation: with all due respect, climate change does not change *everything*.[6] Where conceptually problematic invocations of a knowable objective reality have been effectively challenged, their intellectual status is not redeemed by our appreciation of the evidence of global warming or by our frustration at the denial of that evidence. Likewise, important challenges to realist views of scientific authority are not made irrelevant by our appreciation of the efforts of climate scientists or by our awareness of campaigns to discredit those efforts.

The above observation evidently needs emphasis in view of persistent claims to the contrary: the claim, for example, that a robust philosophical realism is needed to counter the "alternative facts" spun out by political strategists, or that a "postmodern" critique of objectivism has led to a public mistrust of scientific knowledge.[7] Dubious claims of these kinds should, I think, be countered and rejected. But we may also acknowledge the intellectual and pragmatic, including political, perplexities they reflect regarding the realities of, among other things, climate change.

Some of those perplexities were given vivid expression by Bruno Latour in the essay mentioned above, "Why Has Critique Run Out of Steam?: From Matters of Fact to Matters of Concern." Published in 2004, it continues to be widely cited and remains influential in relation to these issues. In a passage of particular interest, Latour quotes a political strategist who, seeking to deflect concern about global warming, advises making "the lack of scientific certainty" a primary point in arguments. "Do you see why I am worried?" Latour asks. "I myself have spent some time in the past trying to show '*the lack of scientific certainty*' inherent in the construction of fact[s]" (227). He continues with a cascade of further questions:

> Was I foolishly mistaken? … Was I wrong to participate in the invention of this field known as science studies? … Why does it burn my tongue to say that global warming is a fact whether you like it or not? Why can't I simply say that the argument is closed for good? … Should we apologize for having been wrong all along? Or should we rather bring the sword of criticism to criticism itself and do a bit of soul searching? … Nothing guarantees, after all, that we should be right all the time … Isn't this what criticism intended to say: that there is no sure ground anywhere? But what does it mean when this lack of sure ground is taken away from us by the worst possible fellows as an argument against the things we cherish?
>
> *(227)*

Here and elsewhere in the essay, Latour poses crucial questions and says powerful things. But, here as elsewhere, the ironies are complex, the allusions ambiguous, and the recommendations somewhat elusive or equivocal.

His aim, Latour writes, is to raise some issues about "critique": "not," he continues, "to depress the reader, but to press ahead, to redirect our meager capacities as fast as possible" (226). But both the redirection proposed and the meaning of the contrast at the heart of it—from "matters of fact" to "matters of concern"—are open to widely differing interpretations. One interpretation, current in the literary academy, is that scholars should turn from critique to more positive activities: to *caring* about texts and other things; to taking writers at their word instead of "digging deeply" into their texts for hidden meanings; to being less concerned with facts and more with affects; and so forth. This interpretation is reflected in an article titled "The New Modesty in Literary Criticism." Its author, Jeffrey Williams, writes as follows:

> In the theory years, you were what your reading was—Marxist, feminist, deconstructionist, queer … Some who were in that camp, with its suspicious habits, began moving away from it by the late 1990s and early 2000s. … Other scholars also began to worry about the effects of theory. In a much-discussed

> … essay, "Why Has Critique Run Out of Steam?," the sociologist of science Bruno Latour worried that *the relativism and social constructionism of postmodern theory had discredited good science.*
>
> *(Williams 2015, emphasis added)*

That is not, of course, quite what Latour said was worrying him.

I suggest another way to interpret Latour's worries below but would offer a word, first, on my perspective on these issues. It derives in part from my association with the literary academy but is shaped more significantly by an intellectual sympathy with the constructivist tradition in philosophy and the social sciences. The interest and sympathy were reflected in talks and articles I presented to academic audiences in the 1990s. At the time, the American academy was much occupied by the "science wars"—fought, it should be recalled, by those defending traditional, largely rationalist–realist views of science against those promoting the relatively novel, largely constructivist views emerging from science studies. I was myself often occupied at the time both with expounding the latter views and with attempting to untangle ignorant confusions and malicious conflations regarding them: for example, their identification as (or with) "deconstructionism," "postmodernism," or "social constructionism" and their conflation with Marxist ideology critique and/or (sometimes in the same breath) classic idealism. Those controversies offered lessons about advocacy under conditions of extreme polarization and motivated distortion. These are, of course, just the conditions that we face today with regard to climate change and that Latour evokes in the opening pages of his essay. We may return, then, to what he probably was and was not saying there.

In urging a turn from "matters of fact" to "matters of concern," Latour was certainly not saying that we should revive triumphalist narratives of the history of science or realist–rationalist explanations of the authority of scientific knowledge. Nor was he repudiating either the critiques of such accounts elaborated over the twentieth century by constructivist theorists of science or the alternative accounts developed in the field of science studies by, among others, himself. Such critiques have continued to win his endorsement, and the alternative accounts continue to figure centrally in his work, including his writings on climate change.[8] What he may have been doing in the essay, at least in part, was calling on his fellow researchers and theorists in science studies to turn from critical treatments of traditional epistemology (or "matters of fact"), which had lost their urgency ("run out of steam"), to more concrete and pressing problems (or "matters of concern"), such as controversies over climate change, where their expertise might be useful. Such calls for increased engagement with current problems were already familiar in the field and remain current in it, along with debates over scholars' commitment to neutrality versus explicit advocacy (Demeritt 2006, Zuiderent-Jerak and Jensen 2007). This interpretation of Latour's essay lacks the drama and scope implied by its title and by various allusions in the text.[9] It has the advantage, however, of making sense in relation to his work as a theorist and practitioner of science studies. It also allows us to have some grip both on the "worries" he evokes rather obliquely in that opening passage and on the responses to such worries that he seems to suggest there.

The fact that "the worst possible fellows" say things that sound like what we say certainly presents problems, but they are not new ones. Indeed, they are ancient and proverbial. The devil or assorted devils have always quoted scripture for their own purposes. That has never been thought a good reason, however, to stop preaching, much less to abandon scripture. It *has* been thought a good reason to clarify one's teachings and to pay close attention to the situations and idioms of those one seeks to address. Constructivist accounts of science are, of course, not scriptural. As it happens, though, this solution to problems of misappropriation is one to which Latour himself turns after detailing his corresponding worries, as a quasi-theologian, about questionable treatments of the New Testament. The title of the work in which he spells this out is *Rejoicing: Or the Torments of Religious Speech*. We may recall his "burn[ing]" "tongue" in the passage quoted above from "Why Has Critique Lost its Steam?" More to the point here, it is what Latour attempts, under the label "diplomacy," in regard to, among other things, constructivist accounts of science in the Anthropocene (Latour 2015).

II

I turn now to a second set of perplexities related to climate change: not efforts by assorted devils to discredit evidence of its reality but the denial or disregard of that evidence on the part of large segments of the public. Current writings by environmentalists invoke a suite of psychological tendencies to explain these reactions—or, as one witty title has it, *Don't Even Think About It: Why Our Brains are Wired to Ignore Climate Change* (Marshall 2014).[10] A different angle on those reactions, and on beliefs and denials more generally, is suggested by an important line of empirically informed inquiry in cognitive theory and philosophy of mind. For those pursuing this line of thought, the question is not, as in classical epistemology, how we, or the philosophers or scientists among us, can arrive at knowledge of the truth. The question is, rather, how humans generally come to believe what they believe or know what they think they know.

In most traditional accounts, beliefs are conceived as discrete, correct-or-incorrect propositions about the world ("the earth is flat," "the earth is round," and so forth) located in our minds. They might be better conceived, however, as dynamic systems of linked assumptions, ideas, images, and recollections and related perceptual and behavioral dispositions. This alternative view is suggested by several intellectual traditions. One is research and theory in empirical psychology, cognitive science, and neuroscience. A second, drawing on the first and on work in theoretical biology, is an influential approach in philosophy of mind.[11] In each of these fields, the relevant accounts stress the dynamic, interactive, and embodied character of the processes and products of cognition: that is, of learning, remembering, figuring out, and planning and of knowledge, memories, ideas, and beliefs. Increasingly in these fields, human cognition is identified not with so-called higher, rational, or computational processes occurring inside our heads but, rather, with our ongoing responsive interactions with our environments as embodied creatures.

Throughout our lives, we interact with our environments in ways that continuously modify our bodily structures and how they operate; and these modifications affect our subsequent interactions with our environments, both what we perceive and how we act. It is not only that our bodily structures and their operations define what we can "detect" about the world, but also that the world that each of us can *act on* and *be acted upon by* is a specific perceptual and behavioral niche. Through the very process of living, moving about in, and exploring the world, we, like other creatures, continuously form cognitive constructs of the features of those niches; and, in turn, we, like other creatures, are continuously shaped—bodily and thus also cognitively—by the features of the niches with which we interact.

Given the reciprocal dynamics just described, we can never become pure spectators of the world, observing a reality altogether independent of us. This is not to say, however, that we are *locked out* of "the real world." It is to say, rather, that, like other creatures, we are inextricably *interlocked with* the real-as-can-be world. The objective reality posited in realist and neorealist ontologies is commonly evoked as "out there," a realm of being distinct and apart from the humans who, as it were, just happen to inhabit it. Our *operative* realities, however, that is, the perceptual and behavioral niches with which we can and do interact and which can be consequential for us, are not distinct from the humans—philosophers and scientists included—who inhabit them. On the contrary, they are continuously realized, brought forth and made real for us, through the very processes of our living in them.

Much of what we experience as external to ourselves and speak of as "reality" can be understood as the relatively stable, provisionally workable configuration of objects, forces, and processes that we continuously construct through our ongoing, effective-enough interactions with our experiential environments. The strong sense we may have of the simple out-there-ness of "what's out there" can be seen, accordingly, not as an undeniable intuition of an ontological given but as the thoroughly contingent product of complex processes of perceptual and behavioral coordination. These processes tend to issue in relatively similar patterns of response to relatively recurrent and stable conditions in our experience, and they are thus relatively predictable and reliable in their effects. We may appreciate, then, the experiential sources and apparent experiential confirmation of classic realist ideas. But we need not endorse those ideas in their classic versions. Specifically, we need not presume any isomorphism or other form of matching or mirroring between, on the one hand, what we experience as, and name, "reality" and, on the other hand, the putatively objective properties of a presumptively independent universe. And we need not presume that the effectiveness of our individual actions or our communally developed technologies depends on our having—in our minds, in our science textbooks, or in our engineering manuals—correct representations of a universe altogether apart from us.

The dynamic-enactivist view of human cognition outlined above was developed relatively independently of the tradition of constructivist epistemology associated with the history and sociology of science. Both, however, reject classic realist, rationalist, and representationalist accounts of knowledge and the alternative accounts of knowledge and belief developed by each are mutually supportive. One of the earliest articulations of constructivist epistemology is Ludwik Fleck's classic work, *Genesis and Development of a Scientific Fact* (1979 [1935]). A scholar of medical history and shrewd amateur sociologist as well as a practicing microbiologist, Fleck sought to describe the formation of scientific knowledge in relation to the more general dynamics of human cognition at both the individual and the sociohistorical levels. Of particular interest here is his analysis of what he calls "belief systems" and what he describes as their "tendency to inertia." He writes: "Once a structurally complete and closed system of beliefs consisting of many details and relations has been formed, it offers tenacious resistance to anything that contradicts it" (27, translation modified). Fleck relates this tendency to what he describes as the ongoing "mutual attunement" of beliefs, perceptions, and material practices among the members of an epistemic community or, in his term, "thought collective." In his account, collectives, through this ongoing process, develop characteristic "thought styles," which is to say, shared habits of perception, interpretation, and explanation.

Several features of belief systems, so understood, are relevant to climate skepticism and denialism. One is *the dense interdependence of their elements*. For example, different sets of communally shared assumptions might dispose us to see—perceive and interpret—a severe drought as evidence of global warming, or as part of a normal weather cycle, or as punishment by the gods. And, reciprocally, our interpreting a drought in one or another of these ways would reinforce those assumptions and, with them, our continued disposition to perceive and respond to comparable events in just those ways. This ongoing positive feedback loop gives considerable stability to belief systems. By the same token, however, it makes those systems strongly resistant to change.

A second relevant feature of belief systems, so understood, is *their social constitution and institutional maintenance*: that is, the fact that they are both formed and stabilized through our ongoing interactions with, among other things, other people, especially members of our particular epistemic communities. A good part of what restabilizes a conviction (e.g., a belief that the earth is flat) in the face of destabilizing experiences (e.g., some clever fellow's demonstration that the earth is round) is the renewal of the collective social practices through which that conviction was formed in the first place—practices, it must be stressed, that are not only verbal but embodied and usually also materially consequential as well. If a belief system remains stable among the members of some collective, it is because the perceptual *and* behavioral dispositions associated with it allow its members to operate effectively in whatever domains of thought and practice are significant in their lives. Thus, convictions or beliefs that are clearly mistaken or irrational in relation to otherwise well-established understandings of the world—for example, the idea that the earth is flat or that droughts are divine punishments—may nevertheless serve, or have served, the members of a particular community well enough in the only worlds they actually inhabit/ed.

This perspective on belief systems allows us to appreciate some aspects of belief perseverance not always recognized by those who invoke the tendency to explain stubborn skepticisms regarding climate change. First, the tendency in question is not a flaw in our mental makeup or a maladaptive hangover from our evolutionary history. Rather, it is a powerful but ambivalently operating—sometimes good, sometimes bad—feature of human cognition. Cognitive plasticity is obviously necessary for any creature that survives, as humans do, by its ability to learn and to modify its behavior responsively. But so also is the countertendency that I call "cognitive conservatism": that is, our ability and inclination to retain our beliefs beyond the occasion of their formation (Smith 1997, 50–51, 84–86). Much of intellectual life (and perhaps of intellectual history) can be understood as the interplay of these complementary human tendencies and their assessment from different perspectives. Thus, given our well-developed capacity to learn from others, we can be duly "informed" and "enlightened" or, as we say, usually of other people, "duped" or "indoctrinated." And, given our ability and inclination to hold fast to what we believe, we can be stubborn in our attachment to error or, as we say, usually of ourselves, steadfast in our defense of truth.

I would also stress, accordingly, the *generality* of cognitive conservatism. The tendencies associated with it—rationalization, dissonance-avoidance, confirmation bias, and so forth—are not restricted to the naive or the uneducated. Physicians hold fast to their diagnoses, scholars to their glosses. Philosophers discredit the intellectual competence of challengers; scientists discount anomalies as flukes. These tendencies are routinely attributed to climate skeptics and climate denialists. We should not be surprised if they were found among environmentalists as well.

III

This last possibility recalls Latour's suggestion, in the long passage quoted above, that those engaged in critique should examine their own attitudes and practices. "Or should we rather bring the sword of criticism to criticism itself and do a bit of soul searching?," he asks and adds: "Nothing guarantees, after all, that we should be right all the time." The suggestion has been taken as urging a wholesale abandonment of practices of critique and has spawned, accordingly, a good deal of rather indiscriminate anti-critical declaiming to that effect.[12] Latour has his own quarrels with some types of critique and often writes—indiscriminately enough—of "the critical spirit" or "the critical mind." But the soul-searching he recommends here can be understood differently: not as critical suicide but, rather, as a type of principled reflexivity urged and often pursued in science studies. Latour himself has urged and pursued it under the term "symmetrical anthropology." Thus he admonishes field anthropologists to "come home" and examine Western, modern beliefs the way they examine the beliefs of other cultures.[13]

A recent "ontological turn" in anthropology is of related interest. The development is usually described as a determination by ethnographers and theoretical anthropologists both to "take seriously" the cosmologies of indigenous peoples and to question seriously their own ontological assumptions.[14] There may be no direct historical connections, but one can see this self-disciplining turn as a radicalization of the more general commitment to explanatory symmetry and reflexivity put forth in the sociology of science in the 1970s and sustained more generally in constructivist studies of science and technology (STS). The aim of those twin commitments was to avoid the commonly lopsided—self-flattering and strongly presentist—explanations of scientific achievement familiar in rationalist epistemology and triumphalist history of science. Although the practices associated with those commitments, particularly with symmetry, are certainly relativistic in some respects, they are neither foolish nor, accurately described and duly understood, properly seen as objectionable. On the contrary, such practices can be valuable in dealing with a range of conceptual and ethical perplexities, including the kind that concern us here.

A study by social psychologist Arlie Russell Hochschild (2016) offers an example relevant to climate change. Hochschild sought to understand why people living in some of the most petro-polluted regions of the United States vote regularly for candidates who support both the oil-extracting, chemical-producing industries that are responsible for the pollution and also the free-market practices that have contributed to the destruction of those voters' natural environment and currently threaten their homes and health. In the study, titled *Strangers in their Own Land*, she calls it "The Great Paradox." Focusing on a coastal region in Louisiana, Hochschild interviewed a good number of such people, following them in their daily routines and speaking with them in their homes, churches, and workplaces. She describes the study as involving efforts to surmount "the wall of empathy" that separated her, a self-described liberal and progressive, from the men and women—all professed members of the conservative Tea Party—whose lives and views she documented: factory workers, fishermen, gospel singers, small-business owners, local officials, and so forth. Hochschild's determination to report their views impartially is evident throughout the book. Of particular interest here are her efforts to find counterparts, in her own sentiments, to those expressed by her subjects and also to find parallels to them in feelings, views, and values commonly expressed by her liberal friends and associates. I return below to some of the Tea Party members' views and sentiments. What I want to note here are both the symmetry and the reflexivity displayed in those efforts: forms of intellectual good practice that can bring us, as they evidently brought Hochschild herself, to insights not otherwise readily available and to understandings of ourselves as well as others that can be politically as well as psychically valuable.[15]

Recent thought about these issues is inevitably shadowed by the coincident political dramas unfolding in the United States. The realities evoked in Hochschild's book are especially haunting. Many oppositional slogans, "Love trumps hate," "We have nothing to fear but fear," and so forth, are probably beside the point. The most politically significant views that Hochschild encountered among her Tea Party subjects were born not of fear or raw hatred but, rather, of nostalgia and deep resentment. While she does not use the latter term, her reports, like those of many other observers, suggest that, along with the sense of a better world lost, a sense of injured merit—and, in effect, of injustice—may be among the most powerful motivators of current rural and working-class politics in the United States.[16] The values and attitudes that Hochschild describes among the people she studied are largely the homely ones familiar to observers and social historians of the American South and Midwest: personal independence, in-group loyalty, hard work, and pride in status. Louisiana Tea Party voters are not resentful of their petro-industry employers, to whom they are grateful for jobs and for whatever measure of local prosperity they see. Nor are they resentful of superrich businesspeople, whom they admire for their industriousness and success. They are resentful of people and groups whom they see as getting special attention or getting ahead unfairly—largely blacks, gays, and immigrants—and perhaps most deeply of those, "liberals" and "elites," whom they see as putting those people and groups ahead while condemning *them* as backward, bigoted, or worse. The Great Paradox is hardly paradoxical at all.

Some reflections suggest themselves accordingly. It is probably not the "postmodern" critique of realist epistemology that makes rural, working-class voters in the United States ignore climate science. It may not even require industry-funded campaigns to win and sustain their support for anti-regulation candidates. It may be enough that they see environmentalism, not altogether incorrectly, as a liberal-elite agenda and that global warming remains, for them, a distant and abstract concept.[17] At least two important lessons seem to follow. One is that hopes for a grassroots movement of resistance to the extraction of fossil fuels must be checked against the force of such views and the force of the political realities to which they give rise. Here as elsewhere, malice can trump prudence and resentment can trump everything. The other is that, to be effective, environmentalist arguments, like arguments against the devil quoting scripture, must be attuned to the conceptual idioms of their actual audiences and calibrated to their operative realities.

IV

In describing, above, the generality of belief-persistence, I noted that the tendencies associated with it might be found among environmentalists as well as among climate skeptics and denialists. What I had in mind was the persistent hope or conviction, common among the writers cited here, that widespread recognition of the perils involved in global warming will lead to the emergence of a climate-conscious and climate-conscientious human collective, either a major new coalition or, in effect, an in-group of humanity as a whole.

Those voicing such hopes often invoke past cultural shifts and large-scale movements as precedents. Thus appeals are made to abolitionism, feminism, the anti-Vietnam War movement, and the acceptance of gay marriage, with the French or Russian revolutions and perhaps, for some, the rise of Christianity as shadow models of historical success.[18] But such appeals must ignore or downplay the significantly different forces and circumstances that exist here: notably, the diffuseness, complexity, and relative impersonality of the harms involved, the magnitude of the immediate costs or sacrifices required, and the uncertainty and abstractness of the benefits, if any, to be gained. In a recent book, *The Great Derangement*, Indian writer Amitav Ghosh, noting a range of discouraging historical precedents and comparably discouraging current political and economic conditions, writes (with a perhaps unwitting double emphasis): "The refusal to acknowledge these realities sometimes lends an air of unreality to discussions of climate change" (Ghosh 2016, 145).[19]

Other differences are equally important. The varied phenomena that we sum up as climate change have aspects that make due recognition both difficult to arrive at and difficult to communicate. Some of the relevant phenomena are violent and manifest but local and transient; others are large and widespread but slow and not readily perceptible; all are exceedingly unevenly distributed globally and even regionally. The relevant information is technical, complex, and not easily absorbed or retained by individuals, even those with specialized training. Such information is also neither ready to hand nor easy to come by, and much of the data is, in fact, disputed among experts. Those in the past who theorized or summoned large-scale uprisings did not confront and probably could not have imagined the combined aggregating, stupefying, and disaggregating effects of the hyper-mass media, which, even as they enable the mobilization of multitudes, sharpen and polarize group differences and isolate communities from potentially illuminating interactions with others.

As emphasized by environmentalists, there is a virtual consensus of informed scientists regarding the destabilizing, destructive effects of anthropogenic climate change, and one could claim a virtual consensus of academics, intellectuals, and thoughtful, well-intentioned people regarding the need for a set of appropriate responses. This statement of double consensus, however, obscures a number of sticky issues. What responses *are* appropriate? By what methods, in what assemblies, and with what authority, will that appropriateness be determined, and in accord with which—or whose—interests, views, and priorities? Latour's essay, "Why has Critique Run out of Steam?," is often invoked, here as elsewhere, as urging us (presumably thoughtful, well-meaning academics and intellectuals) to turn from "matters of fact," meaning, perhaps (among other things) the detailed data of climate science, to "matters of concern," meaning, perhaps (among other things), the actual quotidian threats involved. Given, however, significant differences of situation as well as multiple, divergent interests among us (certainly among humans generally and even just among thoughtful, well-meaning academics and intellectuals), it is not clear, especially when equally justifiable interests conflict, which matters of concern should concern us most. *Humanity?* Every one of us? If not, then who? *The biosphere?* Every nook, niche, and creature? If not, then which ones? *Future generations?* How far into the future? If not thousands of years, then how many?

It is also not clear what form responses can take that would be on a scale great enough to have the required or desired effects. Fossil fuels should, of course, be replaced by renewable sources of energy. But where, when, by what technical, political, and/or economic means, and through what social agencies? Are geo-engineered protections or mitigations (cement pilings, floodgates, cloud seeding, and so forth) altogether to be disdained, as suggested by those intent on purely socioeconomic and/or spiritual transformations? If so, then when exactly should we give up driving cars and having children, and through what form of incentive and/or compulsion, administered and monitored by what governmental agencies? Duly detailed attention to such questions might better prepare us for the conflicts and losses thereby indicated and perhaps encourage the compromises and sacrifices required by all for the survival of anything.[20] Conversely, without acknowledgment of these and other comparably difficult issues, environmentalists' summons to responsible action becomes only too readily ignorable.

Many scientists and environmental activists write as if geophysical facts will triumph of themselves or will do so when the machinations of "merchants of doubt" are exposed. Often overlooked is the significance of the variable conditions of the *reception* of those facts and the significance, for different listeners, of how the accuracy of that information is authorized and how the exposures are mediated. Historians and political scientists, especially as spurred by climate change, have recently joined sociologists of science in noting the limits of the so-called "deficit model" of science communication, which tends to portray scientists as inherently credible deliverers of accurate information to an ignorant public (Howe 2014, Simis et al. 2016, Jasanoff and Simmet 2017). Environmental activists sometimes suggest that the internet or social media amount to vastly expanded, presumably democratic, public spheres. But, as noted by most media scholars and informed commentators, those electronic venues have tended to become more or less tightly sealed social enclosures, as, of course, more familiar concrete spaces (pubs, squares, and town halls included) also could be. The "public sphere" of Critical Theory was always an idealized concept.

Writers voicing hopes for the emergence of an environment-conscious global collective also appeal to allegedly universal human values or natural impulses as dependable motivators. Thus British environmental activist George Marshall writes of the universality of parental care for children and other "nonnegotiable sacred values" (Marshall 2014), and American environmental historian Jedidiah Purdy describes the "self-restraint" that follows from a recognition of personal "guilt" and thus "responsibility" (Purdy 2016).[21] Others write of a "treasuring" of the natural world and life-forms awakened by encounters with natural beauty and sharpened by art or the due study of biology (Wilson and Kellert 1993). The universality and naturalness of such impulses may be doubted, however, along with the political effectiveness of such recognitions and treasurings as may, indeed, occur among various, but probably quite limited, groups of people. Environmentalists invoking such allegedly endemic responses and moral norms, parental love included, must disregard their well-attested cultural variability also their evident failure to prevent persistent patterns of rather nasty behavior among humans everywhere. I remarked above that climate change doesn't change *everything*. It doesn't change *everyone*, either.

The difficulties involved in such appeals and arguments are illustrated in a recent book, *Defiant Earth* (2017), by Australian environmental ethicist Clive Hamilton. Hamilton calls for "a new anthropocentrism" anchored in a recognition of humans' unmatched achievements ("marvels of intellect and culture") joined with our equally unrivalled power to destroy. He goes on to argue that because, unlike other creatures, humans have the freedom to choose, we have the duty to repair or avert the environmental damages and threats caused by the unbridled exercise of our powers. Hamilton describes these ideas as "the defining truths of the age" (55), but his highly abstract moral logic is not likely to transform the behavior of many people. At the same time, the theologically tinged image of humans as demigods obscures an arguably more useful recognition of a range of relevant human attributes, including some reliable, though strictly sublunary, creative capacities.

In addition to the forms of denialism noted above, there is an inclination among environmentalist writers to suggest that the failings of dominant views and values (notably Western, modern, materialistic, and male-associated) are exposed by global warming and will be duly humbled while the benefits of marginal views and values (notably non-Western, traditional, spiritual, and female-associated) are redeemed by it and will be duly elevated.[22] There is sometimes a grim satisfaction to these suggestions of reversals. One could call it Anthropocene *Schadenfreude*: all may go down in darkness, but the despised and rejected—those who suffered under capitalism and imperialism, who railed against secularism and modernity, and who predicted catastrophe and ruin—will, as one writer puts it, have "the last laugh."[23]

Futurology is a high-risk sport for amateurs and must be, for professionals, an increasingly thankless task. It is exceedingly hard to predict how the complex and highly differentiated effects of climate change will play out in the long run or even in the middle run. As I have suggested, there is good reason to doubt that those effects will be countered by the united efforts of a duly enlightened global human community. What seems more likely is the ad hoc emergence of local coalitions in the face of more or less urgent local threats. Barbarism may not be the only alternative to socialism, as Rosa Luxemburg believed and as Isabelle Stengers (2015) echoes. But it also seems likely that, as critical resources (habitable shelter, cultivatable land, food, water, energy, and so forth) become increasingly scarce, there will be struggles, more or less barbarous, between different groups.

My aim here has certainly not been to discredit the efforts of environmental advocates and activists. I admire greatly the scholars and writers whose works I have been citing and would want to share their most optimistic views. My aim, rather, has been to articulate some of the problems I see associated with these efforts and to suggest some alternative ways to respond to the concerns we share. It comes down, perhaps, to advising more relevantly informed imaginings and goals and to suggesting that we grant not, of course, the equal validity of all views or the equal consequentiality of all realities but the equal *force* of all views and realities for those who hold and inhabit them and, therefore, their substantial claim on the attention of those of us who—on whatever terrain, intellectual or practical, ethical or political—must operate in relation to them.

It is just as well that climate change does not change everything or everyone. Adaptive as well as maladaptive capacities will no doubt continue be displayed. At least some past wisdom is likely to remain a resource in the face of large-scale historical obliviousness. Even as realities change radically, humans will probably continue to form collectives, continue to instruct and learn from one another, and, we may hope, operate in many places more or less effectively—and, sometimes, less rather than more barbarously.

Notes

1 See http://media.mcgill.ca/en/content/climate-realism-international-colloquium.
2 For discussion of the elusive meanings of "relativism" and the complex intellectual dynamics of its invocation and disavowal, see Smith 1988, 150–184; 1997, 1–22, 73–87; 2002; 2006, 18–45; 2007; 2011; and 2018, 7–23.
3 In a subsequent work, Kitcher defends a "modest realism" that incorporates, somewhat equivocally, more innovative components (Kitcher 2002).
4 For an engaging image of such an imagined glimpse, see the late nineteenth-century Flammarion woodcut (figure 1 [originally published in *L'Atmosphère: Météorologie Populaire*, Paris 1888, 163]). Here a pilgrim evidently catches a view of the machinery behind the everyday world of trees and houses; day and night; sun, moon, and stars. Notably, aside from one hand, only his head ventures—or, perhaps, can gain admission—into that realm.
5 For comparable views developed in science studies (STS), see Pickering 1995 and Gad et al. 2015.
6 I allude here to Klein 2014 (*This Changes Everything: Capitalism vs. The Climate*).
7 For the former, see, for example, M. P. Lynch 2017; for the latter, see, for example, Manda 2003 and Lepore 2017.
8 See, for example, Latour 2013 *(An Inquiry into Modes of Existence)* and 2017.
9 For example, it does not explain Latour's caustic allusions to deconstruction or to the training of students in the humanities. One notes, however, that the essay was originally a lecture given at the Stanford Humanities Center.
10 See also Hoffman 2015 and Stoknes 2015.
11 For works exemplifying these views and approaches, see Gibson 1979, Varela et al. 1991, Thelen and Smith 1994, Port and van Gelder 1995, Núñez and Freeman 1999, Noë 2004, Chemero 2009, Thompson 2010, Anderson 2014, and Hutto and Myin 2017. Barrett 2011 offers an original, engaging introduction to them.
12 For incisive commentary, see Barnwell 2016.
13 In an important essay (Latour 2010), he illustrates the practice by describing the facts constructed and accorded authority in Western science in pointed parallel with the fetishes constructed and worshiped in the past by Gold Coast natives. For extended discussion of the essay and Latour's efforts to negotiate the relations between science and religion, see Smith 2016.
14 For examples, see Charbonnier et al. 2017. For informed commentary, see Jensen 2017.

15 Such insights and understandings are not, of course, thereby guaranteed. See M. Lynch 2000 for a cautionary analysis of the contingent operations of methodological reflexivity.
16 The sentiment described here is evidently powerful across the current political and demographic spectrum, and not only in the United States. We seem generally to see it as a sense of injustice when displayed by those we ourselves see as victimized and as resentment when we are skeptical of the merit, injury, or idea of justice implicitly invoked.
17 Most of Hochschild's subjects see global warming as a remote problem. Some of them expect a different apocalypse, with a Rapture of the pious, in their own lifetimes.
18 See, for example, Klein 2014, Marshall 2014, Oreskes and Conway 2014, Stengers 2015, and Purdy 2016.
19 For comparable observations and extended analyses, see Clark 2015 and Lorimer 2017.
20 On the significance of varied perspectives on such issues and the lack of public forums for their discussion and negotiation, see Jasanoff and Simmet 2017.
21 Environment-conscious writer Amitav Ghosh suggests that a broad-based religious or quasi-religious movement might supply the needed transformative energy, but he does not appear to anticipate one (Ghosh 2016). Other climate activists are explicit in invoking theology, religion, or religiosity as a desirable or crucial element of a due awakening (see, e.g., Stephenson 2015). For a detailed account of invocations and covert appropriations of religiosity by environmentalists, see Nelson 2010.
22 The image of existing class hierarchies overturned when conditions of survival change radically is evidently a recurrent conceit. See, for example, Pierre Marivaux's *The Island of Slaves* and J. M. Barrie's *The Admirable Crichton,* in both of which, after a shipwreck, practical-minded, able-bodied slaves or servants become the masters.
23 Such people, including writers of fantasy and science fiction—genres disdained by the literary establishment in favor of classically "realist" novels—are what Amitav Ghosh calls, at one point, "the losers." Alluding to several such works by non-Western writers ignored by Western critics but especially relevant to climate change, he comments: "But once again, the last laugh goes to that sly critic, the Anthropocene, which has muddied, and perhaps even reversed, our understanding of what it means to be 'advanced'" (Ghosh 2016, 80).

References

Anderson, Michael L. *After Phrenology: Neural Reuse and the Interactive Brain.* Cambridge, MA: MIT Press, 2014.

Barnwell, Ashley. "Entanglements of Evidence in the Turn Against Critique." *Cultural Studies,* vol. 30, no. 6, 2016, pp. 906–925.

Barrett, Louise. *Beyond the Brain: How Body and Environment Shape Animal and Human Minds.* Princeton, NJ: Princeton University Press, 2011.

Charbonnier, Pierre, Gildas Salmon, and Peter Skafish, editors. *Comparative Metaphysics: Ontology after Anthropology.* London: Rowman and Littlefield, 2017.

Chemero, Anthony. *Radical Embodied Cognitive Science.* Cambridge, MA: MIT Press, 2009.

Clark, Timothy. *Ecocriticism on the Edge: The Anthropocene as a Threshold Concept.* London: Bloomsbury, 2015.

Demeritt, David. "Science studies, climate change and the prospects for constructivist critique." *Economy and Society,* vol. 35, no. 3, 2006, pp. 453–479.

Fleck, Ludwik. *Genesis and Development of a Scientific Fact*. Edited by Thaddeus J. Trenn and Robert K. Merton, translated by Fred Bradley and Thaddeus J. Trenn, University of Chicago Press, 1979 [originally *Entstehung und Entwicklund einer wissenschaftlichen Tatsache*, Basel, 1935].

Gad, Christopher, Casper Bruun Jensen, and Brit Ross Winthereik. "Practical Ontology: Worlds in STS and Anthropology." *Nature Culture* 3, 2015, pp. 67–86.

Ghosh, Amitav. *The Great Derangement: Climate Change and the Unthinkable*. Chicago, IL: University of Chicago Press, 2016.

Gibson, J. J. *The Ecological Approach to Visual Perception*. London: Houghton Mifflin, 1979.

Hamilton, Clive. *Defiant Earth: The Fate of Humans in the Anthropocene*. Cambridge, UK: Polity, 2017.

Hochschild, Arlie Russell. *Strangers in Their Own Land: Anger and Mourning on the American Right*. New York, NY: The New Press, 2016.

Hoffman, Andrew J. *How the Culture Shapes the Climate Change Debate*. Redwood City, CA: Stanford University Press, 2015.

Howe, Joshua P. *Behind the Curve: Science and the Politics of Global Warming*. Seattle, WA: University of Washington Press, 2014.

Hutto, Daniel D., and Eric Myin. *Evolving Enactivism: Basic Minds Meet Content*. Cambridge, MA: MIT Press, 2017.

Jasanoff, Sheila, and Hilton R. Simmet. "No funeral bells: Public reason in a 'post-truth' age." *Social Studies of Science*, vol. 47, no. 5, 2017, pp. 751–770.

Jensen, Casper Bruun. "New Ontologies? Reflections on Some Recent 'Turns' in STS, Anthropology and Philosophy." *Social Anthropology*, vol. 25, no. 4, 2017, pp. 525–545.

Kitcher, Philip. *The Advancement of Science: Science without Legend, Objectivity without Illusions*. New York, NY: Oxford University Press, 1995.

———. *Science, Truth and Democracy*. New York, NY: Oxford University Press, 2002.

Klein, Naomi. *This Changes Everything: Capitalism vs. The Climate*. Toronto, ON: Simon & Schuster, 2014.

Latour, Bruno. "Why Has Critique Run Out of Steam? From Matters of Fact to Matters of Concern." *Critical Inquiry*, vol. 30, no. 2, 2004, pp. 225–248.

———. "On the Cult of the Factish Gods." *On the Modern Cult of the Factish Gods*, translated by Catherine Porter and Heather MacLean, Durham, NC: Duke University Press, 2010, pp. 1–66.

———. *An Inquiry into Modes of Existence: An Anthropology of the Modern*. Translated by Catherine Porter, Cambridge, MA: Harvard University Press, 2013.

———. "Diplomacy in the Face of Gaia: Bruno Latour in Conversation with Heather Davis." In *Art in the Anthropocene*, edited by Heather Davis and Etienne Turpin, London, UK: Open Humanities Press, 2015, pp. 43–58.

———. *Facing Gaia: Eight Lectures on the New Climatic Regime*. Cambridge, MA: Polity, 2017.

Lepore, Jill. "Autumn of the Atom: How Arguments about Nuclear Weapons Shaped the Debate over Global Warming." *The New Yorker*, vol. 92, no. 47, January 30, 2017, p. 22.

Lorimer, Jamie. "The Anthropo-Scene: A Guide for the Perplexed." *Social Studies of Science*, vol. 47, no. 1, 2017, pp. 117–142.

Lynch, Michael. "Against Reflexivity as an Academic Virtue and Source of Privileged Knowledge." *Theory, Culture and Society*, vol. 17, no. 3, 2000, pp. 26–54.

Lynch, Michael P. "Kick This Rock: Climate Change and Our Common Reality." *The New York Times*, June 5, 2017.

Manda, Neera. *Prophets Looking Backward: Postmodern Critiques of Science and Hindu Nationalism in India.* New Brunswick, NJ: Rutgers University Press, 2003.

Marshall, George. *Don't Even Think About It: Why Our Brains are Wired to Ignore Climate Change.* London, UK: Bloomsbury Publishing, 2014.

Nelson, Robert H. *The New Holy Wars: Economic Religion versus Environmental Religion in Contemporary America.* University Park, PA: Penn State University Press, 2010.

Noë, Alva. *Action in Perception.* Cambridge, MA: MIT Press, 2004.

Núñez Rafael, and Walter J. Freeman, editors. *Reclaiming Cognition: The Primacy of Action, Intention, and Emotion.* Exeter, UK: Imprint Academic, 1999.

Oreskes, Naomi and Eric M. Conway. *The Collapse of Western Civilization: A View from the Future.* New York, NY: Columbia University Press, 2014.

Pickering, Andrew. *The Mangle of Practice: Time, Agency, and Science.* Chicago, IL: University of Chicago Press, 1995.

Port, Robert F., and Timothy van Gelder, editors. *Mind as Motion: Explorations in the Dynamics of Cognition.* Cambridge, MA: MIT Press, 1995.

Purdy, Jedidiah. *After Nature: A Politics for the Anthropocene.* Cambridge, MA: Harvard University Press, 2016.

Simis, Molly J., Haley Madden, and Michael A. Cacciatore. "The Lure of Rationality: Why does the Deficit Model Persist in Science Communication?" *Public Understanding of Science,* 2016, vol. 25, no. 4, pp. 400–414.

Smith, Barbara Herrnstein. *Contingencies of Value: Alternative Perspectives for Critical Theory.* Cambridge, MA: Harvard University Press, 1988.

———. *Belief and Resistance: Dynamics of Contemporary Intellectual Controversy.* Cambridge, MA: Harvard University Press, 1997.

———. "Reply to an Analytic Philosopher." *The South Atlantic Quarterly*, vol. 101, no.1, 2002, pp. 229–242.

———. *Scandalous Knowledge: Science, Truth and the Human.* Edinburgh: University of Edinburgh Press, 2005/ Durham, NC: Duke University Press, 2006.

———"The Chimera of Relativism: A Tragicomedy." *Common Knowledge,* vol. 17, no. 1, 2011, pp. 13–26.

———. "Anthropotheology: Latour Speaking Religiously." *New Literary History*, vol. 47, nos. 2–3, 2016, pp. 331–351.

———. *Practicing Relativism in the Anthropocene: On Science, Belief, and the Humanities.* London, UK: Open Humanities Press, 2018.

Stanford, P. Kyle. *Exceeding Our Grasp: Science, History, and the Problem of Unconceived Alternatives.* New York, NY: Oxford University Press, 2006.

———. "Naturalism without Scientism." In *The Blackwell Companion to Naturalism*, edited by Kelly James Clark, Hoboken, NJ: John Wiley & Sons, 2016, pp. 91–108.

Stengers, Isabelle. *In Catastrophic Times: Resisting the Coming Barbarism.* Translated by Andrew Goffey, London, UK: Open Humanities Press, 2015.

Stephenson, Wen. *What We're Fighting for Now Is Each Other: Dispatches from the Front Lines of Climate Justice.* Boston, MA: Beacon, 2015.

Stoknes, Per Espen. *What We Think About (When We Try Not to Think About) Global Warming: Toward a New Psychology of Climate Action.* Hartford, VE: Chelsea Green Publishing, 2015.

Thelen, Esther, and L. B. Smith. *A Dynamic Systems Approach to the Development of Cognition and Action.* Cambridge, MA: MIT Press, 1994.

Thompson, Evan. *Mind in Life: Biology, Phenomenology, and the Sciences of Mind.* Cambridge, MA: Harvard University Press, 2010.

Varela, Francisco J., Evan Thompson, and Eleanor Rosch. *The Embodied Mind: Cognitive Science and Human Experience*. Cambridge, MA: MIT Press, 1991.

Williams, Jeffrey J. "The New Modesty in Literary Criticism." *The Chronicle of Higher Education*, January 5, 2015.

Wilson, Edward O., and Stephen R. Kellert, editors. *The Biophilia Hypothesis*. Washington, DC: Island Press, 1993.

Zuiderent-Jerak, Teun, and Casper Bruun Jensen. "Unpacking 'Intervention' in Science and Technology Studies." *Science as Culture*, vol. 16, no. 3, 2007, pp. 227–235.

INDEX

Note: Page numbers in *italics* indicate figures on the corresponding pages.

www.ingramcontent.com/pod-product-compliance
Lightning Source LLC
LaVergne TN
LVHW021130300926
849160LV00010B/940
9781138370043